The New Geography of Jobs

The New Geography of Jobs

◆

ENRICO MORETTI

MARINER BOOKS
HOUGHTON MIFFLIN HARCOURT
Boston   New York

First Mariner Books edition 2013

www.hmhco.com

Library of Congress Cataloging-in-Publication Data
Moretti, Enrico.
The new geography of jobs / Enrico Moretti.
p. cm.
Includes bibliographical references and index.
ISBN 978-0-547-75011-8   ISBN 978-0-544-02805-0 (pbk.)
1. Labor market—United States. 2. Economic development—United
States. 3. Equality—United States. 4. Technological innovations—
Economic aspects—United States. I. Title.
HD5706.M596 2012
331.10973—dc23   2012007933

Printed in the United States of America
DOC 10 9 8 7 6
4500655327

To Ilaria

CONTENTS

The New Geography of Jobs

INTRODUCTION

MENLO PARK IS a lively community in the heart of Silicon Valley, just minutes from Stanford University's manicured campus and many of the Valley's most dynamic high-tech companies. Surrounded by some of the wealthiest zip codes in California, its streets are lined with an eclectic mix of midcentury ranch houses side by side with newly built mini-mansions and low-rise apartment buildings. In 1969, David Breedlove was a young engineer with a beautiful wife and a house in Menlo Park. They were expecting their first child. Breedlove liked his job and had even turned down an offer from Hewlett-Packard, the iconic high-tech giant in the Valley. Nevertheless, he was considering leaving Menlo Park to move to a medium-sized town called Visalia. About a three-hour drive from Menlo Park, Visalia sits on a flat, dry plain in the heart of the agricultural San Joaquin Valley. Its residential neighborhoods have the typical feel of many Southern California communities, with wide streets lined with one-story houses, lawns with shrubs and palm trees, and the occasional backyard pool. It's hot in the summer, with a typical maximum temperature in July of ninety-four degrees, and cold in the winter.

Breedlove liked the idea of moving to a more rural community with less pollution, a shorter commute, and safer schools. Menlo Park, like many urban areas at the time, did not seem to be heading in the right direction. In the end, Breedlove quit his job, sold the Silicon Valley house, packed, and moved the family to Visalia. He was not the only one. Many well-educated professionals at the time were leaving cities and moving to smaller communities because they thought those communities were better places to raise families. But things did not turn out exactly as they expected.

In 1969, both Menlo Park and Visalia had a mix of residents with a wide range of income levels. Visalia was predominantly a farming community with a large population of laborers but also a sizable number of professional, middle-class families. Menlo Park had a largely middle-class population but also a significant number of working-class and low-income households. The two cities were not identical—the typical resident of Menlo Park was somewhat better educated than the typical resident of Visalia and earned a slightly higher salary—but the differences were relatively small. In the late 1960s, the two cities had schools of comparable quality and similar crime rates, although Menlo Park had a higher incidence of violent crime, especially aggravated assault. The natural surroundings in both places were attractive. While Menlo Park was close to the Pacific Ocean beaches, Visalia was near the Sierra Nevada range and Sequoia and Kings Canyon National Parks.

Today the two places could not be more different, but not in the way David Breedlove envisioned. The Silicon Valley region has grown into the most important innovation hub in the world. Jobs abound, and the average salary of its residents is the second highest in America. Its crime rate is low, its school districts are among the best in the state, and the air quality is excellent. Fully half of its residents have a college degree, and many have a PhD, making it the fifth best educated urban area in the nation. Menlo Park keeps attracting small and large high-tech employers, including most recently the new Facebook headquarters.

By contrast, Visalia has the second lowest percentage of college-educated workers in the country, almost no residents with a postgraduate degree, and one of the lowest average salaries in America. It is the only major city in the Central Valley that does not have a four-year college. Its crime rate is high, and its schools, structurally unable to cope with the vast number of non-English-speaking students, are among the worst in California. Visalia also consistently ranks among American cities with the worst pollution, especially in the summer, when the heat, traffic, and fumes from farm machines create the third highest level of ozone in the nation.

Not only are the two communities different, but they are growing more and more different every year. For the past thirty years, Silicon Valley has been a magnet for good jobs and skilled workers from all over the world. The percentage of college graduates has increased by two-thirds, the second largest gain among American metropolitan areas. By contrast, few high-paying jobs have been created in Visalia, and the percentage of local workers with a college degree has barely changed in thirty years—one of the worst performances in the country.

For someone like David Breedlove, a highly educated professional with solid career options, choosing Visalia over Menlo Park was a perfectly reasonable decision in 1969. Today it would be almost unthinkable. Although only 200 miles separate these two cities, they might as well be on two different planets.

The divergence of Menlo Park and Visalia is not an isolated case. It reflects a broader national trend. America's new economic map shows growing differences, not just between people but between communities. A handful of cities with the "right" industries and a solid base of human capital keep attracting good employers and offering high wages, while those at the other extreme, cities with the "wrong" industries and a limited human capital base, are stuck with dead-end jobs and low average wages. This divide—I will call it the Great Divergence—has its origins in the 1980s, when American cities started to be increasingly defined by their residents' levels of education. Cities with many college-educated workers started attracting even more, and cities with a less educated workforce started losing ground. While in 1969 Visalia did have a small professional middle class, today its residents, especially those who moved there recently, are overwhelmingly unskilled. Menlo Park had many low-income families in 1969, but today most of its new residents have a college degree or a master's degree and a middle- to upper-class income. Geographically, American workers are increasingly sorting along educational lines. At the same time that American communities are desegregating racially, they are becoming more segregated in terms of schooling and earnings.

Certainly any country has communities with more or less educated residents. But today the difference among communities in the United States is bigger than it has been in a century. The divergence in educational levels is causing an equally large divergence in labor productivity and therefore salaries. Workers in cities at the top of the list make about two to three times more than identical workers in cities at the bottom, and the gap keeps growing.

Cities with a high percentage of skilled workers offer high wages not just because they have many college-educated residents and these residents earn high wages. This would be interesting but hardly surprising. But something deeper is going on. A worker's education has an effect not just on his own salary but on the entire community around him. The presence of many college-educated residents changes the local economy in profound ways, affecting both the kinds of jobs available and the productivity of every worker who lives there, including the less skilled. This results in high wages not just for skilled workers but for most workers.

I consider the Great Divergence to be one of the most important developments in the United States over the past thirty years. As we will discover, the growing economic divide between American communities is not an accident but the inevitable result of deep-seated economic forces. More than traditional industries, the knowledge economy has an inherent tendency toward geographical agglomeration. In this context, initial advantages matter, and the future depends heavily on the past. The success of a city fosters more success, as communities that can attract skilled workers and good jobs tend to attract even more. Communities that fail to attract skilled workers lose further ground.

The growing divergence of American communities is important not just in itself but because of what it means for American society. While the divide is first and foremost economic, it is now beginning to affect cultural identity, health, family stability, and even politics. The sorting of highly educated Americans into some communities and less educated Americans into others tends to magnify and exacerbate all other socioeconomic differences. For example, there are vast differences in life expectancy among inhabitants of American cities, and these differences have been expanding for the past three decades. The divorce rates, crime rates, and political clout of different communities have also been diverging. These trends are reshaping the very fabric of our society.

The United States is not in particularly high spirits these days. Fear of economic decline is widespread, and insecurity about America's standing in the world and its economic future is growing. Talk of

the "death of the American dream" is everywhere, from well-artic-
ulated op-ed pieces to crude talk radio shows, from casual barber-
shop conversations to highbrow academic symposia. In a nation
sharply divided along political lines, concern about the economy is
shared almost equally by those on the left and on the right.

On the surface it seems we have good reason to be worried.
Middle-class salaries are declining. Good jobs are scarce. Take the
typical forty-year-old male worker with a high school education:
today his hourly wage is 8 percent lower than his father's was in
1980, adjusted for inflation. This means that for the first time in recent American history, the average worker has not experienced an
improvement in standard of living compared to the previous generation. In fact he is worse off by almost every measure. On top of
this, income inequality is widening. Uncertainty about the future
is now endemic.

But the economic picture is more complex, more interesting,
and more surprising than the current debate suggests. America's
labor market is undergoing a momentous shift. While some sectors and occupations are dying, others are growing stronger, and
still others, just born, promise to alter the landscape dramatically.
Most of all, the geography of jobs is changing in profound and irreversible ways. While these trends are national, even global, in
scope, their effects are profoundly different in different cities and
regions of the country. For example, the effects of globalization,
technological progress, and immigration on American workers are
not uniform across the United States. They favor the residents of
some cities and hurt the residents of others. As old manufacturing
capitals disappear, new innovation hubs are rising and are poised
to become the new engines of prosperity. An unprecedented redistribution of jobs, population, and wealth is under way in America,
and it's likely to accelerate in the decades to come.

Some of the changes in the economic map reflect long-run forces that are outside our control. Others can be shaped and managed. But none of them are random, chaotic, or unpredictable. In the end, they all reflect clear and rather basic economic principles. Unfortunately, they tend to be obscured by the flood of data on the fluctuations of the stock market or the latest employment numbers. The focus on short-term events often results in information that is incomplete, irrelevant, or both. What happened today, this week, or even this month is not very illuminating, because the fundamentals of an economy evolve at a much slower pace.

But if we take a step back and look at the big picture, the forces that have been driving these changes reveal themselves very clearly. They are far more fascinating and much more important than the daily movements of the Dow Jones. This book examines the long-term trends that really matter to our lives—the vast changes that have taken place in the American labor market over the past three decades and the economic forces underlying these changes. But it also looks forward, seeking to provide insight into the trends that will shape our economy over the next three decades.

Economists like to distinguish *cyclical* change, the ups and downs of the economy driven by the endless cycle of recessions and expansions, from *secular* change, the long-run developments that are driven by deep-seated but slower-moving economic dynamics. Most of the current public debate on the economy—in the media, in Congress, in the White House—focuses on the former. The time horizon in this debate is six months or a year at most: How do we end the recession? What should be in this year's budget? How will unemployment affect the next election? In this book, the focus is almost entirely on the forces that drive long-run trends. Understanding why these changes are taking place, where they are occurring, and how they are affecting individual Americans is crucial. Our jobs, our communities, and our economic destiny are at stake.

The changes taking place in the United States can be seen around the globe. New economic powerhouses are displacing old ones. What used to be tiny, barely visible dots on the map have turned into thriving megalopolises with thousands of new companies and millions of new jobs. Nowhere are these changes more obvious than in the Chinese city of Shenzhen. If you have not heard of it, you will. It is one of the fastest-growing cities in the world. In just three decades it has gone from being a small fishing village to being a huge metropolis with more than 10 million residents. In the United States, a fast-growing city like Las Vegas or Phoenix may triple or quadruple in size over a thirty-year period. Shenzhen's population has grown by more than *300 times* in the same period. In the process, Shenzhen has become one of the manufacturing capitals of the world.

Shenzhen's rise is truly remarkable because it parallels almost perfectly the decline of U.S. manufacturing centers. Thirty years ago Shenzhen was an unremarkable small town that no one outside of southern Guangdong Province had even heard of. Its fate—as well as the fate of millions of American manufacturing workers —was sealed in 1979, when the Chinese leadership singled it out to be the first of China's "Special Economic Zones." These zones quickly became a magnet for foreign investment. In turn, that flow of investment led to thousands of new factories. These factories are where many American manufacturing jobs have gone.

As Detroit and Cleveland have declined, Shenzhen has grown. Massive production facilities of all kinds carpet the region. Every year the skyline adds new high-rise offices and apartments, and its workforce swells as more and more farmers leave rural areas to look for better-paying jobs in its cavernous factories. The Chinese call it the city with "one high-rise a day and one boulevard every three days." As you walk along its wide streets, you feel the city's energy and optimism. Shenzhen has been China's top exporter for the past two decades and has built one of the world's busiest ports, a sprawling facility dotted with huge cranes, enormous trucks, and containers of all colors. Twenty-four hours a day, seven days a week, 365 days a year, these containers are loaded onto enormous cargo ships bound for the West Coast of the United States. Twenty-five million of these containers leave the port each year, almost one per second. In less than two weeks that merchandise will be on a truck headed for a Walmart distribution center, an IKEA warehouse, or an Apple store.

Shenzhen is where the iPhone is assembled. If there is a poster child of globalization, it is the iPhone. Apple has given as much attention to designing and optimizing its supply chain as to the design of the phone itself. The process by which the iPhone is produced illustrates how the new global economy is reshaping the location of jobs and presenting new challenges for American workers.

Apple engineers in Cupertino, California, conceived and designed the iPhone. This is the only phase of the production process that takes place entirely in the United States. It involves product design, software development, product management, marketing, and other high-value functions. At this stage, labor costs are not the main consideration. Rather, the important elements are creativity and ingenuity. The iPhone's electronic parts — sophisticated, but not as innovative as its design — are made mostly in Singapore and Taiwan. Only a few components are made in the United States. The last phase of production is the most labor-intensive: workers assemble the hardware and prepare it for shipping. This part, where the key factor is labor costs, takes place on the outskirts of Shenzhen. The facility is one of the largest in the world, and its sheer size is extraordinary: with 400,000 workers, dormitories, stores, and even cinemas, it is more like a city within a city than a factory. If you buy an iPhone online, it is shipped directly to you from Shenzhen. Incredibly, when it reaches the American consumer, only one American worker has physically touched the final product: the UPS delivery guy.

At a superficial level, the story of the iPhone is troubling. Here you have an iconic American product that has captivated consumers everywhere, but American workers are involved only in the initial innovation phase. The rest of the process, including the making of the sophisticated electronic components, has been moved overseas. It is therefore natural to wonder what might be left to American workers in the decades to come. Is America entering a phase of irreversible decline?

Over the past half century, the United States has shifted from an economy centered on producing physical goods to one centered on innovation and knowledge. Jobs in the innovation sector have been growing disproportionately fast. The key ingredient in these jobs is human capital, which consists of people's skills and ingenuity. In other words, humans are the essential input—they are coming up with the new ideas. The same two forces that have decimated traditional manufacturing, globalization and technological progress, are now driving the rise of jobs in the innovation sector. The Great Recession has temporarily halted this growth, but the long-term trend points upward.

Globalization and technological progress have turned many physical goods into cheap commodities but have raised the economic return on human capital and innovation. For the first time in history, the factor that is scarce is not physical capital but creativity. Not surprisingly, innovators capture the largest share of the value of new products. The iPhone is made up of 634 components. The value created in Shenzhen is very low, because assembly can be done anywhere in the world. Even sophisticated electronic parts, like flash memories and retina displays, create limited value, because of strong global competition. The majority of the iPhone's value comes from the original idea, its unique engineering, and its beautiful industrial design. Essentially this is why Apple receives $321 for each iPhone—much more than any part supplier involved in physical production. This matters tremendously, not just for Apple's profit margin and for our sense of national pride, but because it means good jobs.

The innovation sector includes advanced manufacturing (such as designing iPhones or iPads), information technology, life sciences, medical devices, robotics, new materials, and nanotechnology. But innovation is not limited to high technology. Any job that generates new ideas and new products qualifies. There are entertainment innovators, environmental innovators, even financial innovators. What they all have in common is that they create things the world has never seen before. We tend to think of innovations as physical goods, but they can also be services—for example, new ways of reaching consumers or new ways of spending our free time. Today this is where the real money is. A part of the $321 that Apple receives ends up in the pockets of Apple's stockholders, but some of it goes to Apple's employees in Cupertino. And because of the company's great profitability, it has the incentive to keep innovating and to keep hiring workers. Studies show that the more innovative a company is, the better paid its employees are.

You might think that the rise of innovation is pretty exciting if you work for, say, Google or a biotech company but that it doesn't matter all that much if you're a teacher or a doctor or a police officer. After all, the majority of Americans will never work for a high-tech startup. Why should they care about the rise of innovation? As it turns out, however, innovation matters not only for the well-educated workers who are directly employed by high-tech firms—the scientists, engineers, and creators of new ideas—but for most American workers.

If you take a walk in one of America's cities, most of the people you see on the street will be store clerks and hairstylists, lawyers and waiters, not innovators. About a third of Americans work either for the government or in the education and health services sectors, which include teachers, doctors, and nurses. Another quarter are in retail, leisure, and hospitality, which includes people working in stores, restaurants, movie theaters, and hotels. An additional 14 percent are employed in professional and business services, which include employees of law, architecture, and management firms. In total, two-thirds of American jobs are in the local service sector, and that number has been quietly growing for the past fifty years. Most industrialized nations have a similar percentage of local service jobs. The goods and services in this sector are locally produced and locally consumed and therefore do not face global competition. Although jobs in local services constitute the vast majority of jobs, they are the *effect*, not the *cause*, of economic growth. One reason is that productivity in local services tends not to change much over time. It takes the same amount of labor to cut your hair, wait on a table, drive a bus, or teach math as it did fifty years ago. By contrast, productivity in the innovation sector increases steadily every year, thanks to technological progress. In the long run, a society cannot experience salary growth without significant productivity growth. Fifty years ago, manufacturing was the driver of this growth, the one sector responsible for raising the wages of American workers, including local service workers. Today the innovation sector is the driver. Thus, what happens to the innovation sector determines the salary of many Americans, whether they work in innovation or not.

A second reason that the rise of innovation matters to all of us has to do with the almost magical economics of job creation. Innovative industries bring "good jobs" and high salaries to the communities where they cluster, and their impact on the local economy is much deeper than their direct effect. Attracting a scientist or a software engineer to a city triggers a *multiplier effect*, increasing employment and salaries for those who provide local services. In essence, from the point of view of a city, a high-tech job is more than a job. Indeed, my research shows that for each new high-tech job in a city, five additional jobs are ultimately created outside of the high-tech sector in that city, both in skilled occupations (lawyers, teachers, nurses) and in unskilled ones (waiters, hairdressers, carpenters). For each new software designer hired at Twitter in San Francisco, there are five new job openings for baristas, personal trainers, doctors, and taxi drivers in the community. While innovation will never be responsible for the majority of jobs in the United States, it has a disproportionate effect on the economy of American communities. Most sectors have a multiplier effect, but the innovation sector has the largest multiplier of all: about three times larger than that of manufacturing. Later we will discover why this is the case. For now, let me just point out that the multiplier effect has important and surprising implications for local development strategies. One is that the best way for a city or state to generate jobs for less skilled workers is to attract high-tech companies that hire highly skilled ones.

But this does not mean that everyone wins in the new innovation economy. The shifts in the labor market over the past three decades have created an economic map that is inherently uneven. As the Great Divergence accelerates, American communities are becoming increasingly different from one another. We're used to thinking of the United States in dichotomous terms: red versus blue, black versus white, haves versus have-nots. Today there are *three* Americas. At one extreme are the brain hubs—cities with a well-educated labor force and a strong innovation sector. They are growing, adding good jobs and attracting even more skilled workers. At the other extreme are cities once dominated by traditional manufacturing, which are declining rapidly, losing jobs and residents. In the middle are a number of cities that could go either way. The three Americas are growing apart at an accelerating rate. The paradox is that the very success of the country's engine of growth is generating a deep and growing inequality among American communities. As we will discover, the winners and losers in this process are not always the people we expect.

It wasn't supposed to be this way. At the peak of the dot-com frenzy in 2000, observers of all stripes almost unanimously concluded that "the New Economy gives both companies and workers more locational freedom." In *The World Is Flat*, one of the most influential books about globalization, Thomas Friedman famously argued that cell phones, e-mail, and the Internet lowered communication barriers so much that location was irrelevant. Distance was dead. Geography didn't matter.

This argument has continued to resonate. The idea is that no matter where people live, they can share knowledge and move products at virtually no cost. According to this view, the good jobs, now concentrated in high-cost locations such as Silicon Valley and Boston, will quickly disperse to low-cost locations, both in the United States and abroad. An experienced software engineer in India makes $35,000. The same person in Silicon Valley makes $140,000. Why would U.S. firms keep hiring in Silicon Valley when they could save so much by outsourcing? By the same token, if labor costs are three times higher in Silicon Valley than in Mobile, Alabama, companies will eventually relocate to Alabama. This process of dispersion, the argument goes, will be faster than the dispersion of manufacturing jobs, because moving software codes across DSL lines is easier than moving bulky goods across borders. In this vision of the future, the great innovation hubs of America will disappear from the map and innovation jobs will disperse evenly across the country. The key prediction of this view is the convergence of American communities. Low-cost areas will attract more and more of the new, high-paying jobs. Cities that have been lagging behind—the Clevelands, the Topekas, and the Mobiles—will grow much faster. Bogged down by their high costs, San Francisco, New York, Seattle, and similar cities will decline.

But the data don't support this view. In fact, the opposite has been happening. In innovation, a company's success depends on more than just the quality of its workers—it also depends on the entire ecosystem that surrounds it. This is important, because it makes it harder to delocalize innovation than traditional manufacturing. A textile factory is a stand-alone entity that can be put pretty much anywhere in the world where labor is abundant. By contrast, a biotech lab is harder to export, because you would have to move not just one company but an entire ecosystem.

A growing body of research suggests that cities are not just a collection of individuals but complex, interrelated environments that foster the generation of new ideas and new ways of doing business. For example, social interactions among workers tend to generate learning opportunities that enhance innovation and productivity. Being around smart people makes us smarter and more innovative. By clustering near each other, innovators foster each other's creative spirit and become more successful. Thus, once a city attracts some innovative workers and innovative companies, its economy changes in ways that make it even more attractive to other innovators. In the end, this is what is causing the Great Divergence among American communities, as some cities experience an increased concentration of good jobs, talent, and investment and others are in free fall. It is a trend that is reshaping not just our economy but our entire society in profound ways. It implies that a growing part of inequality in America reflects not just a class divide but a geographical divide.

This does not mean there is no merit to the view that low-cost areas are destined to catch up. At a global level, the most important economic development of the past decade is the incredible improvement in the standard of living in developing nations such as Brazil, China, Poland, Turkey, India, and even some African countries. Their strong economic performance has greatly reduced the gap between these countries and rich nations, thus contributing to a marked convergence in income levels. This is welcome news. Although seldom recognized, inequality has plummeted when measured at a global level. The catch-up experienced by the American South over the past fifty years is another example of convergence. Many southern states were significantly poorer than the rest of the country in the 1960s but grew more rapidly in the following decades.

Yet in both cases the process of catching up was geographically uneven. Some southern cities—Austin, Atlanta, Durham, Washington, D.C., Dallas, and Houston, for example—grew much faster than others, thus increasing the disparity among communities in the South. Developing countries exhibit similar regional differences. In China, Shanghai has reached a per capita income close to that of a rich nation. Its students outperform American and European students in standardized tests by a wide margin. Its public infrastructure is better than that of most American cities. But agricultural communities in western China have made much less progress. The regional differences in China have clearly grown, even if the difference between China and richer countries has shrunk.

What is driving these trends? Why is divergence increasing and not decreasing, as many people expected? This book explores the root causes of these trends and what they mean for American communities, examining the special features of the United States that make it attractive for innovative companies. Understanding why jobs in this sector are clustering in America's innovation hubs is the key to understanding our economic future.

Despite all the hype about the "death of distance" and the "flat world," where you live matters more than ever. Whether you work inside or outside the innovation sector, whether you are self-employed or work for others, where you live greatly affects all aspects of your life, from your career to your finances, from the kind of people you meet to the values your children are exposed to. As America's cities grow apart, it is more important than ever to understand the new geography of jobs.

The coming chapters are a voyage through the new economic landscape. We will explore cities that are rising and cities that are dying. We will travel to faraway lands and familiar backyards. We will meet a color scientist at Pixar and a bookbinder in San Francisco. We'll walk the streets of Seattle's up-and-coming Pioneer Square, once known for its methadone clinics and now home to companies such as Zynga and Blue Nile. We will visit Berlin, Europe's sexiest city but still surprisingly poor, and Raleigh-Durham, which is relatively dull but increasingly prosperous. We will discover why Walmart.com had to leave Arkansas and what this means for auto workers in Flint and call-center employees in Albuquerque. In doing so, we will discover how the changes in the world economy are reshaping the American workplace and American communities. These are the forces that will determine the location of future jobs and the fate of particular cities and regions. We will learn what causes these changes and how they will affect our careers, our communities, and our way of life.

But first we need to understand how we got here. The United States used to be poor and then became a prosperous middle-class society in the last century. How? And can we maintain our prosperity, even in the midst of tumultuous change?

1

◆

AMERICAN RUST

EVERY YEAR, MILLIONS of Chinese and Indian farmers leave their villages and move to sprawling urban centers to work in an ever-growing number of factories. Americans can't help but observe, with a mix of awe and anxiety, the millions of manufacturing jobs created in those cavernous facilities, the constant flow of goods coming out of them, and the remarkable rise in standard of living that follows. Americans may have forgotten, but not too long ago this was us: our transition from a low-income society to a middle-class society used exactly the same engine—good manufacturing jobs.

In 1946, the year after World War II ended, American families were poor by today's standards. Infant mortality was high. Salaries and consumption were low. Household appliances like refrigerators and washing machines were rarities. The purchase of new shoes was a major event for most people. Only 2 percent of households had a TV. But over the next three decades, American society experienced one of the most impressive economic transformations in history. Salaries and incomes grew at an astonishing rate. Consumption exploded at all levels of society. An unprecedented sense of affluence and optimism pervaded most parts of the country. By 1975 infant mortality had been cut in half and living standards had doubled. Household appliances had become so cheap that everyone could afford them. Purchasing new shoes became an unremarkable occurrence, and virtually all households owned a TV. In one short generation, America had turned into a middle-class nation.

During those years, the growth of middle-class incomes was tied to the rising productivity of manufacturing industries such as automobiles, chemicals, and steel. For millions of workers, the stability of a good, well-paid job in a factory was the American Dream. With it came all the perks of middle-class life, both economically and culturally, from home ownership to weekends off and summer vacations. In short, it meant prosperity and optimism. The most dynamic areas in the country at the time were manufacturing meccas like Detroit, Cleveland, Akron, Gary, and Pittsburgh. These cities were the envy of the world. Their might and prosperity were clearly and visibly linked to factories, smokestacks, greasy equipment, and the production of tangible, often heavy physical goods. Detroit reached the peak of its economic power in 1950, when it became the third richest city in the United States. It was the Silicon Valley of its day, thanks to its unprecedented agglomeration of cutting-edge companies, many of which were world leaders in their sectors, and it attracted the most creative innovators and engineers. The identification of America's prosperity with industrialization reached its height in the 1950s, when Charles Wilson, then the CEO of General Motors, famously said, "What is good for General Motors is good for the country, and vice versa."

The engine that made all of this possible was an unprecedented rise in the productivity of workers. Because of better management practices and a tremendous surge in investment in new and more modern machines, an American factory worker in 1975 could produce twice as much for each hour worked as the same worker could in 1946. This productivity gain had two reinforcing effects on America's prosperity. First, it resulted in substantial wage increases. While the increase in productivity was largely confined to manufacturing, the wage increases were not. Additionally, the higher productivity enabled manufacturers to produce goods more efficiently and therefore more cheaply. Goods such as cars and household appliances, which had been prohibitively expensive, became affordable mass commodities. To buy an average-sized Chevrolet, the median household had to spend half of its income in 1946 but less than a quarter in 1975.

This combination of lower prices and higher salaries had a profound effect on the cultural and economic structure of American society and was accompanied by an enormous transformation of the consumer experience, driven by the proliferation of shopping malls and the rise of mass marketing and advertising. Consumption increased so much that a new word was coined: *consumerism*. After centuries of struggle with nature and scarcity, this new opulent society offered the average family an unprecedented level of material well-being. In the typical family, parents expected their children to be twice as well off as they were, just because they lived in America.

In the fall of 1978, manufacturing employment reached its peak, with almost 20 million Americans working in factories. That year Jimmy Carter was president, *Grease* was the top-grossing movie, and the soap opera *Dallas* was transfixing TV viewers of all stripes. The economy did well that fall, logging a solid expansion in both gross domestic product (GDP) and jobs. Then suddenly the engine stopped. Manufacturing employment, the workhorse that had single-handedly pulled America from the uncertainties of the Great Depression to the stability of the postwar years, slowed down, then stopped, then began to move backward. In early 1979 oil prices skyrocketed following the Iranian revolution. The car industry was hit first, but the malaise quickly spread to other sectors as higher production costs led companies to shed jobs. When the price of oil finally came down, job losses continued. What initially looked like a temporary downturn turned into a protracted, painful decline that continues to this day.

The Demise of a Giant

The demise of this engine of growth is truly staggering. Although the U.S. population is now much larger than it was in 1978, there are half as many jobs in manufacturing as there were at its peak. Right now manufacturing jobs are more the exception than the rule, employing fewer than one in ten American workers, a fraction that keeps declining year after year. Today an American is significantly more likely to work in a restaurant than in a factory. Consider Figure 1, which shows the number of manufacturing jobs over the past quarter century. Since 1985 the United States has lost an average of 372,000 manufacturing jobs every year. This does not simply reflect short-term phenomena such as recessions, as manufacturing often loses jobs even during expansions. Nineteen out of twenty sectors that will experience the largest job losses in the next ten years, according to the U.S. Department of Labor, are in manufacturing, led by "Cut and sew apparel manufacturing," "Apparel Knitting Mills," and "Textile and fabric finishing and coating mills." If current trends continue, there will be more laundry workers than manufacturing workers in America when my son, who is now three years old, enters the labor market.

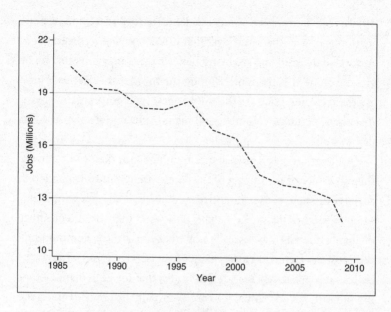

Figure 1. The decline of manufacturing jobs

Manufacturing is no longer the engine of prosperity for local communities. If anything, the opposite is true. The big manufacturing centers of America, once proud and wealthy, have been humbled and are now struggling with a shrinking population and difficult economic prospects. They are pale ghosts of what they used to be, and many are at risk of disappearing from the economic map entirely. Their names are now synonymous with urban blight and irreversible decline. Between the 2000 and 2010 censuses, the metropolitan area that experienced the biggest drop in population was New Orleans, because of Hurricane Katrina. But just below New Orleans came Detroit (minus 25 percent), Cleveland (minus 17 percent), Cincinnati (minus 10 percent), Pittsburgh (minus 8 percent), Toledo (minus 8 percent), and St. Louis (minus 8 percent). It is as if year after year Rust Belt cities keep being hit by their own Hurricane Katrina. Having peaked in size in the late

1950s, Detroit has been losing residents for fifty years, and its population is now at the same level that it was one hundred years ago. A third of its residents are living below the poverty line. Its rate of violent crime is consistently among the highest in America. Gone are the factories, gone are the smokestacks, and gone are the greasy machines. Gone too are the good manufacturing jobs that used to pay so well.

And yet, as staggering as the numbers are, they can't convey the full extent of the impact of the demise of manufacturing on our society. In some sense, an entire way of life is disappearing. One often-missed point in the debate on jobs is that the direct effect of manufacturing job losses is not what hurts these communities the most: as factories close, many of the service jobs in these cities also disappear. My research indicates that for each manufacturing job lost, 1.6 additional jobs are eventually lost outside that sector in affected communities. These losses include barbers, waiters, carpenters, doctors, cleaners, and retailers. The loss of construction jobs is particularly damaging. In Rust Belt communities, construction traditionally represented the best-paying job outside of manufacturing for workers without a lot of schooling. But all those construction jobs were ultimately supported by manufacturing incomes. As those dry up, so do the livelihoods of other members of the community.

The national mood has dipped even lower. There is clear and tangible anxiety about the direction of the country, and it transcends the travails of the 2008–2010 recession. One recent poll found that the decline of U.S. manufacturing is giving Americans a "sense of economic precariousness." The main reason, according to a *Boston Globe* analysis, is the pervasive concern that "the United States no longer makes enough *stuff*." In the 1984 song "My Hometown," Bruce Springsteen perfectly captured the worried mood of countless East Coast and Rust Belt communities hit badly by factory closures. He sings of empty storefronts on Main Street and the jobs that are leaving town for good: "Seems like there ain't nobody wants to come down here no more." Twenty-five years later, the weary feeling in Springsteen's lyrics seems to be even more widespread.

The manufacturing sector's reversal of fortune is one of the most important facts in America's economic history of the past six decades. Much of the pessimism about our future, including the discussion about the end of American exceptionalism, can be traced to the decline of manufacturing. While the standard of living for the average family more than doubled from 1946 to 1978, it has been largely stagnant ever since. Take, for example, the average American worker, a forty-year-old male with a high school diploma and about twenty years of work experience. Between 1946 and 1978, this guy's hourly wage measured in today's dollars went up from $8 to $16. Since 1978 his wage has actually gone *down* by $2.

What happened? What caused such a stunning reversal? Many people place the blame on banks and financiers. This notion is ingrained in the national psyche and considerably older than the Occupy Wall Street movement. In Oliver Stone's acclaimed movie *Wall Street*, the economic transformation of the 1980s is portrayed as a fight between the honesty and purity of Main Street, embodied by a solid, contented blue-collar union representative played by Martin Sheen, and the corruption and moral recklessness of Wall Street, embodied by Martin Sheen's son Charlie. Charlie Sheen plays a young stockbroker willing to do anything to get ahead in the ruthless world of corporate raiders and ends up almost destroying the company where his father works. Thirty years later, Hollywood's view of America's economic woes is unchanged. In the

2010 movie *The Company Men*, Ben Affleck portrays a white-collar employee who loses his job when a greedy CEO orders savage layoffs in an effort to appease Wall Street and boost the company's stock price. The similarities are striking. In both movies the good guys make real, physical things—in the early movie they work for an airline, in the later one they work for a shipbuilding company—while the bad guys scheme with stocks and options, spend their time aggressively shouting buy or sell orders, and end up destroying jobs. In one of the most poignant scenes of *The Company Men*, two of the laid-off workers visit the rusting, abandoned shipyard and muse, "We used to make real things here."

Greedy financiers and pushy yuppies in sleek business suits make compelling villains in any narrative, but the reality is that Wall Street did not kill blue-collar America. History did. The problems with American manufacturing jobs are structural, and they reflect deep economic forces that have been gaining strength over the past half century: globalization and technological progress.

From Factories to Private Schools

If there is a brand that embodies the industrial history of the United States of America, it is Levi Strauss & Company. When I moved to San Francisco in the early 1990s, there was still a Levi Strauss factory in the city. Levi's was founded there in 1853 during the Gold Rush, when a twenty-four-year-old German immigrant began supplying sturdy pants to gold prospectors. The San Francisco factory had been in operation ever since, one of the thousands of manufacturing plants that used to dot American urban neighborhoods. In the summer of 1994 I visited the facility, where dozens of mostly Latin American women were cutting and sewing the trademark 501 jeans. I distinctly remember wondering how long they would be able to hang on. For years the company tried to protect its American employees, but with wages at $9 to $14 per hour plus benefits, its production costs were significantly higher than those of its competitors. Finally, in 2001, the company closed down all of its facilities in the United States and relocated production to Asia. The San Francisco factory is now an elite Quaker private elementary school, with a yearly tuition of $24,045.

I was not surprised. If anything, I found it remarkable that Levi Strauss had resisted outsourcing for so long. Other companies in its sector—Gap, Ralph Lauren, Old Navy—had shifted their production overseas much earlier. In this respect, the apparel sector is typical of the manufacturing sector as a whole. In the decade after World War II, textiles were a major piece of the U.S. labor market. America's most important industrial cluster in terms of jobs was not the Detroit auto industry but the New York garment industry. As recently as the mid-1980s, more than a million American workers were still employed by U.S. companies making clothing and garments. Today that number has dropped by more than 90 percent. Take a minute to check where your clothes were made. If you are wearing clothes sold by an American company, a third-party vendor in a place like Vietnam or Bangladesh probably manufactured them. American brand names are thriving, but only a handful of jobs—in design, marketing, and sales—remain in the United States.

What is interesting is that on a superficial level, this story is similar to that of the iPhone: design and marketing jobs remain in this country but vendors in Asia make all the parts. However, there is a major difference. For apparel—and for traditional production in general—the design and marketing jobs that remain are few and are not growing in any appreciable way, while in the innovation sector the design and engineering jobs are numerous and growing fast.

Until recently we did not import much from low-wage countries. As recently as 1991, these countries accounted for less than 3 percent of U.S. manufacturing imports, a number too small to affect a large number of jobs. But over the past two decades the world has become a global village of ever-expanding commerce. By 2000 the percentage of imports from low-income countries had doubled, and by 2007 it had doubled again, with China accounting for most of the increase. What happened was an enormous shift in the production of physical goods away from rich countries with high labor costs to poorer countries with low labor costs. As the iPhone illustrates, there are much better places on earth to make many physical goods, including fairly sophisticated ones.

Since labor is cheap in developing countries, factories there tend to use fewer machines than in the United States, a fact that gives those factories the additional advantage of being more flexible and more adaptable to sudden change. In a recent interview, an American businessman doing business in China was quoted as saying, "People think China is cheap, but really, it's fast." An American industrial designer who works in China added, "People are the most adaptable machines. Machines need to be reprogrammed. You can have people doing something entirely different next week." Unlike American factories, factories in China can respond almost overnight to changes in production plans or design.

The effect of globalization on American blue-collar jobs is not the same everywhere. An important new study by the economists David Autor, David Dorn, and Gordon Hanson finds that the impact of imports from China depends heavily on where you live. Cities like Providence and Buffalo have manufacturing sectors that are heavily skewed toward traditional, low-value-added productions similar to those of China, and they have experienced large negative effects from the increased competition. By contrast, cities like Washington, D.C., and Houston are engaged in very different kinds of manufacturing and have experienced much smaller effects. In cities that directly compete with China, imports were found to cause rising local unemployment, decreased labor-force participation, and lower local wages. Interestingly, not all of these costs are borne by the workers directly displaced: a part of the cost is borne by other Americans in the form of government aid. The study finds that imports from China increased the use of welfare payments such as unemployment insurance, food stamps, and even disability insurance, which is often used as a hidden form of welfare. In essence, while the direct effect of trade was highly localized, the ultimate costs were at least in part shouldered by taxpayers in the rest of the country through federally funded programs.

The effect of globalization also varies enormously depending on companies' ability to react. A recent study by Nicholas Bloom, Mirko Draca, and John Van Reenen shows that increased trade with developing countries causes faster technological upgrading but that the eventual effect depends on each company's willingness to adjust. Using a comprehensive data set on half a million firms in twelve industrialized countries between 1996 and 2007, the economists found that firms facing Chinese import competition tend to react by upgrading their technology: they buy more computers, spend more on R&D, take out more patents, and update their management policies. The irony is that this external threat has become an important driver of productivity gains for American companies and therefore economic growth for the country. But not everyone gains. While high-tech firms successfully respond to the threat, low-tech firms—those with limited innovation, limited investment in IT, and limited productivity—have a harder time reacting to Chinese imports and end up laying off workers or disappearing. Thus globalization stimulates technical progress, which in turn increases the demand for educated workers, but it reduces the demand for unskilled workers.

The Rise of Manufacturing Hipsters

Of course, there are exceptions to the decline. High-end fashion, for example, is less sensitive to labor costs than other manufacturing sectors and depends more on where designers and skilled tailors may be. Moreover, there is a clear resurgence in the appeal of local artisanal goods. Everything that is locally produced, from food to clothes, bicycles to furniture, is fashionable these days. From New York and Providence to Portland and Minneapolis, an increasing number of artisanal workshops have appeared and are selling goods in trendy local boutiques.

The neighborhood where the San Francisco Levi's factory used to be is now crowded with dozens of workshops offering handcrafted products—outfits like Cut Loose, which makes a line of hip clothing that is sewn and dyed to order in San Francisco. Literally across the street from the Levi's factory, a boutique called The Common specializes in "purveying, producing and designing durable and timeless staples crafted using traditional manufacturing techniques." Their funky ready-to-wear shirts are designed, cut, and sewn in California. A few blocks down the street, what used to be an auto body shop has recently become a full-fledged chocolate factory. Housed in a beautiful red brick building, Dandelion Chocolate sells $9 chocolate bars handmade on the premises by two intense-looking clean-shaven hipsters with a passion for uncovering the best organically grown Madagascar beans. A couple of miles to the east, DODOcase has taken over an old, almost bankrupt bookbindery to handcraft iPad cases complete with a personalized monogram and eco-friendly bamboo parts. In Brooklyn, local food production is booming. "It seems as if every 28-year-old guy in the borough has a line of artisanal pickles," reports *Metropolis* magazine. At the Brooklyn Navy Yard, the workshops of dozens of small manufacturers are bustling with activity. Space is in short supply as more and more local producers want in. The contrast with the hundreds of abandoned factories in Detroit and Flint could not be starker.

A century ago Brooklyn was one of the capitals of urban manufacturing in America. At its peak during World War II, its Navy Yard employed 70,000 people who worked in three separate shifts, twenty-four hours a day. Judging from all the activity today, one might be tempted to conclude that urban manufacturing is back, in the form of small-scale high-tech production that employs college-educated young people and targets local markets. Ferra Designs, which rents a 10,000-square-foot space in the Navy Yard, is a metal shop specializing in architectural fabrication. Partner Jeff Kahn recently told *Metropolis* that most of his fifteen employees are industrial designers with a college degree from the nearby Pratt Institute. "Most of them are under thirty. They're into craftsmanship; they want to know how to build things. It's a renaissance." He thinks the Navy Yard's success signals the potential for a revival of urban manufacturing in America. "The cost of doing business is going up in China," he argues. "This country has an opportunity to regain some of its manufacturing base, using cutting-edge technology and a new generation of interested youth."

It is a mantra that is becoming widespread in many urban areas across the United States, and it is being embraced by thousands of young people interested in working with their hands. When I visited the latest high-end handcrafted cloth maker to open near the Levi's factory, the irony of the situation was inescapable: in the same place where only twenty years ago undereducated Hispanic women used to cut and sew Levi's garments, there are now dozens of overeducated young white people cutting and sewing similar products.

These initiatives are culturally interesting, which is why they are the subject of a growing number of feature articles in local papers, and are worth supporting. Local production helps capture some of the wealth that might otherwise have gone overseas, and in many cases it has a significantly smaller ecological footprint. But it is clear that these initiatives cannot be the answer to the lack of jobs in America. First, they are destined to remain niche phenomena. The number of jobs created is simply too small to make a dent. More fundamentally, these jobs can't be the *driver* of job growth for a community. They will always be the *result* of wealth created in some other sector. This is an important but often misunderstood point. In traditional manufacturing, products tend to be sold everywhere in the world. But consumption of locally produced goods depends by definition on the existing wealth of an area. After all, someone in the local economy has to pay for these $40 handmade T-shirts and $9 artisanal chocolate bars. In the case of cities like New York and San Francisco, financial and high-tech industries are the sectors generating the wealth that supports these local artisanal efforts.

Additionally, an important part of the appeal of local manufacturing is that we perceive it as something special and different. This makes this sector inherently difficult to scale up past the narrow limits of what consumers identify as "unique." Take American Apparel, for example, which operates the largest remaining North American garment factory. The facility employs five thousand workers and is located in a multistory building in downtown Los Angeles, just a few blocks from the high-rises of the financial center. The company's marketing focuses on the fact that it pays workers decent wages—$12 an hour for sewing operators—and offers health insurance. Its T-shirts have become immensely popular among many young, educated, hip, urban consumers. What is remarkable is that there is nothing special about these T-shirts besides where they are made. It is a sign of the times that production of clothes in the United States has become such a rarity that the mere fact of being manufactured in downtown L.A. distinguishes the product, making American Apparel clothing attractive to hipsters in Williamsburg, Austin, and D.C. It is not a bad business model. People perceive the brand as cool, so the company can sell its products at a premium and therefore cover its relatively high production costs. Unfortunately for other textile workers in the United States, however, this model cannot be replicated widely. By definition, the American Apparel model rests on the premise that its products are unique. If everyone were producing in U.S. urban areas, the company would lose its only competitive advantage.

How China and Walmart Helped the Poor

Americans do not suffer from low self-esteem. Unlike European politicians, our politicians routinely refer to the United States as "the greatest country on earth" and to American workers as "the best in the world." There are of course many reasons to be proud of being American. But just like workers in other countries, American workers are good at making some things and not so good at making others. And that's fine. In a global economy, you do not need to excel at everything. In fact, you shouldn't even try. It is much better to let other countries provide things for you, as long as you are good enough to offer something else in exchange. It makes sense: David Beckham should focus on being the best possible soccer player and let others build his house, cut his hair, and make his clothes.

Economists argue about almost everything, but they all agree on the principle of comparative advantage. The key insight is that if each country concentrates on the industries in which it is relatively more productive, everybody wins. Each country exports the product it is particularly good at making and in exchange imports other goods produced relatively more efficiently abroad. The net result is that we all end up a little richer. Actually, we end up a lot richer. America's national income today is billions of dollars higher than it would be without international trade. This part of the globalization promise has worked so well that we now take it for granted. American consumers today expect that most electronic goods—from computers to wall-sized flat-screen TVs—will become cheaper year after year. Evidence shows that prices of consumer goods have fallen most in sectors where imports from China have increased the most.

Interestingly, this trend has benefited the poor more than the rich. Two economists at the University of Chicago recently studied the consumption patterns of households with different incomes. Every time a participant in their study came back from shopping at the grocery store, she scanned all the products purchased. Using this remarkably rich data set, the economists discovered that the price of the goods purchased by the typical low-income consumer tended to increase much less than the price of goods consumed by the typical high-income consumer. Since 1994, the price index for the poorest 20 percent of families has grown three times more slowly than the price index for the richest 20 percent.

There are two plausible explanations: China and Walmart. Low-income consumers tend to buy proportionally more goods that are made in China and other low-wage countries—things like toys, cheap clothes, and affordable consumer electronics. Thanks to globalization, the price of these goods has risen less than others over the past fifteen years, and in many cases, like that of consumer electronics, it has actually declined. By contrast, consumers with higher incomes tend to buy proportionally more personal services—everything from haircuts and housecleaning to restaurant meals and health services. Since personal services are less exposed to foreign competition, high-income consumers end up benefiting less from globalization.

The expansion of superstores such as Walmart has also played a role. Since low-income consumers shop twice as frequently in superstores as high-income consumers, the impact is larger on the poor than on the rich. And even if low-income consumers do not shop at Walmart, they tend to benefit from the price competition it creates, as many of its stores are in low-income areas. The economist Emek Basker studied the impact that the opening of a Walmart superstore had on local prices and found not only that Walmart's prices were lower but also that the entry of a new superstore caused other local stores to lower their prices by 6 to 12 percent.

The principle of comparative advantage tells us that countries with different industrial structures have the most to gain from trading with each other and the least to lose in terms of job losses. Emerging countries such as China, Brazil, and India have economies that are different enough from America's that the gains from trade are potentially large, with job growth in the United States concentrated in the innovation sector. Once you see things from the point of view of comparative advantage, the standard perspective on international competition offered by many in the media starts to look silly. The traditional view is that if one of our trading partners, such as China, becomes more productive, it's terrible news for us, because it means that that country will steal our jobs. But trade is not a zero-sum game like a football match, where if your opponent wins you lose. The reality is that if one of our trading partners becomes more productive, the goods we are buying from that country become cheaper. This makes us—the consumers—a little richer.

Overall, the effect of imports from low-wage countries has been highly uneven, with less skilled workers taking most of the job losses. At the same time, these imports cost less, and that saves consumers money. One of the paradoxes of globalization is that the very people who have been hit the hardest in terms of jobs have gained more as consumers.

The Productivity Paradox

Globalization is only part of the story of the decline of manufacturing jobs. For all its woes, the United States still produces many physical goods. We tend to forget this, because almost everything we pick up in the store says "Made in China." While this is true of many consumer products, it is not true of many high-end nonconsumer goods, such as airplanes, industrial machines, and advanced medical devices. Newspapers rarely report this fact, but American factories produce the same output as China, more than double that of Japan, and several times that of Germany and Korea. The U.S. manufacturing sector alone is larger than the entire British economy, and it is growing. Since 1970, U.S. manufacturing has doubled its output, and it keeps expanding over time.

What is going on here? If production keeps increasing, how come manufacturing jobs keep disappearing? The reason for this apparent contradiction is that thanks to technological improvements and investment in new and more sophisticated machinery, U.S. factories are significantly more efficient than they used to be, so fewer and fewer American workers are needed to produce the same number of goods. Today the average factory worker in the United States makes $180,000 worth of goods each year, more than three times what he produced in 1978. Higher productivity is a very good thing for the economy in general, but the impact on blue-collar jobs is dramatic. Take General Motors, for example. In the 1950s, the glory years of Detroit, each GM employee made on average seven cars per year. By the 1990s that number had risen to about thirteen cars per year, and now it is twenty-eight cars per year. The math of job losses is pretty simple: compared to 1950, today GM needs four times *fewer* workers for each car it produces. Those workers who do have a job in manufacturing are now more productive than before and therefore earn higher wages, but there are far fewer of them.

This is another of the intriguing paradoxes of economic growth: increases in productivity lower prices for consumers and raise wages, but they ultimately end up killing jobs. Critics stress the loss of jobs, but the reality is that an increase in the productivity of labor is the main way in which societies become more prosperous and elevate their standard of living. There is nothing new in this phenomenon. The American economy went through a similar transformation when it moved from being mainly agricultural to being industrialized. One hundred and fifty years ago, half of U.S. workers labored in a field. Today only one percent of workers are in agriculture, and most of us will live our entire lives without meeting a single farmer. Yet owing to technological improvements — things like tractors, fertilizers, better seeds — crops today are produced in much greater quantities and much more cheaply. As agricultural productivity soared in the twentieth century, rural income rose but the need for agricultural workers declined, so farmers moved en masse to urban factories. The same transformation is taking place again, as manufacturing productivity destroys manufacturing jobs but is making us, on average, richer.

New industries are not immune. Take a look at the evolution of American jobs in computer manufacturing and semiconductor manufacturing in Figure 2. Worldwide sales of computers and semiconductors have been exploding over the past twenty-five years, but employment in those two sectors has been plummet-

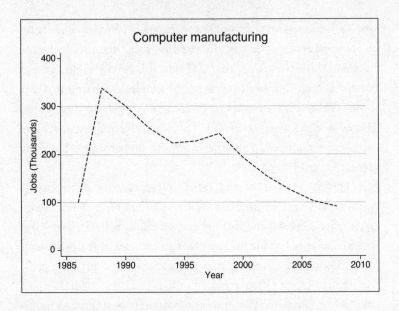

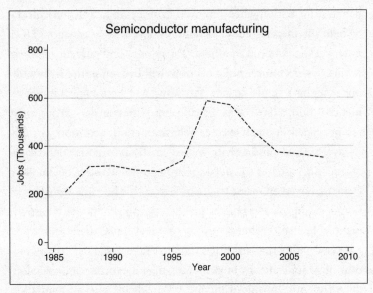

Figure 2. Jobs in computer and semiconductor manufacturing

ing. There are fewer production jobs in the computer manufacturing industry today than there were in 1975, before the personal computer was introduced. Indeed, the figure shows that the top year for employment was 1988—the year that Apple launched the Macintosh IIx and Commodore sold 1.5 million C64s to adoring fans, including me. Laptops were rare, computing power was ridiculously low, and tablets were made of stone. The semiconductor industry tells a similar story. It is common for states and municipalities to fight to attract semiconductor plants, and yet the number of production jobs in the industry has been declining for a decade.

This is probably the aspect of manufacturing's demise in the United States that I find most remarkable: even sophisticated high-tech electronic products have not been immune to the general malaise of the goods-producing sector. When I showed these graphs to Paul Thomas, the chief economist at Intel, the world leader in semiconductors, he was not particularly surprised. Automation has made the production of PCs and semiconductors much less labor-intensive. The parallel with the automobile industry is particularly striking. On top of that, the assembly and manufacture of many of the parts have moved abroad, just as we saw in the case of the iPhone. The first batch of two hundred Apple I computers was assembled by Steve Jobs and Steve Wozniak in Jobs's famous garage in Los Altos in 1976. Production didn't stray far for a few years. During the 1980s, Apple was manufacturing most of its Macs in a factory in Fremont, California. But in 1992 Apple shut down the factory and shifted production first to cheaper parts of California and Colorado, then to Ireland and Singapore. All other American companies followed the model. As James Fallows once put it, "Everyone in America has heard of Dell, Sony, Compaq, HP, Lenovo-IBM ThinkPad, Apple, NEC, Gateway, Toshiba. Almost no one has heard of Quanta, Compal, Inventec, Wistron, Asustek.

Yet nearly 90 percent of laptops and notebooks sold under the famous brand names are actually made by one of these five companies in their factories in mainland China."

Paul Krugman once joked that "depressions, runaway inflation, or civil war can make a country poor, but only productivity growth can make it rich." He was right. Data show that over the past two hundred years, the growth in per capita income of Americans has closely tracked the growth in labor productivity. This is true for every country in the world and for most periods of history. It makes sense. After all, higher labor productivity simply means that a worker can generate more products for each hour worked. Where do these productivity increases come from? Throughout human history, innovation and technological progress have always been the significant drivers of improvement in people's standard of living. Innovation is the engine that has enabled Western economies to grow at unprecedented speed ever since the onset of the industrial revolution. In essence, our material well-being hinges on the continuous creation of new ideas, new technologies, and new products.

The Hollowing Out of the American Labor Market

The effects of globalization and technological change on the labor market have been highly uneven. While the number of blue-collar workers in manufacturing has plummeted since 1978, the number of engineers in manufacturing has doubled. In general, job opportunities in the U.S. labor market as a whole have been concentrated in high-skill, high-wage jobs (professional, technical, and managerial occupations) and low-skill, low-wage jobs (food service, personal care, and security service occupations). Job opportunities for middle-wage, middle-skill white-collar and blue-collar workers have declined sharply. As pointed out by the MIT labor economist David Autor, the labor market is losing its middle. It's hollowing out.

New technologies tend to favor highly skilled workers, reduce the need for many occupations calling for medium-level skills, and barely affect occupations at the low end of the skill spectrum. In an influential 2003 paper, Autor and two colleagues showed that computers and robots are particularly efficient at performing routine tasks—tasks that can be accomplished by following explicit rules, such as repetitive customer service, record-keeping, and many other middle-income white-collar jobs—but they are inefficient at nonroutine tasks. Many tasks that used to be performed by bank tellers, for example, are now performed by ATMs or Web-based programs. But jobs that involve nonroutine tasks—either abstract or manual—have fared quite differently. Occupations such as carpenter, truck driver, housecleaner, security guard, and many others that are defined by nonroutine manual tasks have not been particularly hurt by computers. Jobs that involve "nonroutine problem-solving and complex communications tasks"—those in science, engineering, marketing, and many other areas—have actually been made more productive by computers. For example, a journalist, architect, or scientist is now much more productive because she can use computers and the Internet on her job.

American Pastoral, Philip Roth's great novel about the shifting Zeitgeist of American society in the 1960s and 1970s, chronicles the rise and fall of a Jewish family in New Jersey. But almost as poignant as the unraveling of Seymour "Swede" Levov's personal dreams is Roth's description of the unraveling of Newark's social fabric. In the story, Levov has inherited a small glove factory, a disappearing microcosm of experienced blue-collar workers deeply attached to their jobs, proud of their skills, happy to teach their trade to the next generation. Describing the satisfaction and self-respect that comes from a job well done, Roth contrasts the values of these workers with the values of the uprooted and transient new labor force. The stable middle-class jobs in the glove factory have been replaced with more precarious service jobs that have high turnover and that elicit little pride.

The hollowing out of the labor force and the evaporation of the middle class is not a passing trend, nor is it unique to Newark or the United States—it is common across industrialized economies. Autor has looked at changes in employment since 1993 in sixteen major European Union economies within three broad sets of occupations—low-, middle-, and high-wage. Just as in the United States, the number of middle-wage jobs has declined in all countries while the number of low- and high-wage jobs has increased.

The Tides of History

We spend the best part of our lives at work. Every morning we say goodbye to our loved ones and rush to our offices, cubicles, counters, factories, labs, or whatever place we call "work." For most hours of the day, for most days of the year, and for most years of our lives, our best energies are dedicated to our jobs. Our jobs have become so important that in many cases they define how people perceive us and even how we perceive ourselves. They determine our standard of living and where we live. For some of us, our salary and work schedule determine what sort of family we have, how many children we can afford, and how much time we spend with them. In short, our private and collective well-being depends on what kinds of jobs are out there and what security they might offer. The news on this front has been bad for quite a while.

I am not talking just about the Great Recession of 2008 to 2010. Recessions and economic expansions have always affected the number and kind of jobs available, and the years since 2008 have been particularly tough in this regard. But recessions and ex-

pansions are short-run phenomena. Their ups and downs have always occurred and always will. They are just a small part of the picture. Far more important—and interesting—are the longer-term
trends that ultimately determine our standard of living.

Lately we have seen some signs that the long decline in manufacturing employment might be slowing down. Wages in China
have been creeping up, a predictable effect of increased prosperity. China's move to revalue its currency, the yuan, has further increased labor costs from the perspective of American companies.
General Electric has reopened an appliances factory in Kentucky
after years overseas. Carbonite has brought a call center back to
Boston from India. After twelve years in Mexico, Otis Elevators
is moving its production from Nogales back to South Carolina.
There are even signs of "insourcing," the opposite of outsourcing,
in which foreign companies invest in production facilities in the
United States. A Chinese company called Yuncheng has opened
a factory in Spartanburg, South Carolina, finding it cheaper than
Shanghai. All these signs have generated a growing perception
among pundits and commentators that the U.S. manufacturing
sector is about to turn a page and experience a renaissance. Yet
while all of these examples make compelling media stories, they
are not representative. They capture our attention precisely because they are exceptions that buck the trend.*

The perception of a forthcoming manufacturing comeback
was further bolstered by unexpected gains in manufacturing employment in 2011, the first time in many years that production jobs
grew in a significant way. But the reality is that the gains in 2011,
while certainly welcome news, came after much worse than usual
job losses during the recession years of 2008 and 2009. Looking

---

* In the case of Yuncheng, it turns out that an important factor in the decision to
move to South Carolina was a generous payroll tax credit from the state.

forward, there is little to suggest that the long-run downward trajectory that we saw in Figure 1 is about to change in a permanent way.

When confronted with painful job losses, many people insist that we can and should turn back the clock by protecting the manufacturing sector from all external and internal threats. Some believe that there is something special and uniquely American about producing physical "stuff" and that we should use the full power of the federal government to stop the bleeding of "good blue-collar jobs" from the United States. The more sophisticated version of this argument is articulated in countless op-ed pieces, *New York Times* articles, and a wave of new books that argue that proper legislative action can stop the decline of manufacturing. In this version, proposed policies are often labeled with appealing names like "National Strategy for Manufacturing," but they invariably involve protectionist policies, subsidies, or both.

Essentially, the "manufacturing activists" propose fighting history. Their arguments—both the sophisticated version and its populist sibling—ignore the simple fact that the forces that have caused the decline of manufacturing are difficult to stop. Like King Canute, the English king who believed he could make the tides recede and then almost drowned, the activists cannot simply command the forces of history.

If the jobs are not in manufacturing, then where are they? Right now there are about 141 million workers in the United States. About 112 million work in the private sector, the rest in various forms of government jobs. We collectively put in 2,522,228,000,000 hours of work each year. If we do not make physical things anymore, what do we do all day? More important, what will we do tomorrow? What will propel us forward?

2

◆

SMART LABOR: MICROCHIPS, MOVIES, AND MULTIPLIERS

DOMINIC GLYNN IS a mathematician, and if you think doing math all day is dull, think again. Glynn is a color scientist and lead engineer at Pixar Animation Studios, where he spends his days bringing animated creatures to life. His office, located in Pixar's bright red brick facility in Emeryville, California, overflows with toys, which is not unusual on Pixar's campus. He has worked on many films, including *Cars, Ratatouille, WALL-E, Up,* and *Toy Story 3*. If you liked the colors in those movies, Glynn and his team are the ones to thank. Glynn is in his thirties, plays baroque violin, and has a beautiful three-year-old daughter. When I met him, he was busy finishing *Cars 2*. He told me that the math he uses is simple, but for some reason I was skeptical. Technically, what he does is called image-mastering engineering, which essentially consists of creating mathematical models of human color vision. It is a mix of color science, computer science, and mathematics. He starts with equations and ends up with the amazingly colorful stories that have made Pixar the industry leader.

Pixar's creative genes run deep. The studio was founded by the iconic *Star Wars* director George Lucas and then acquired by Apple's Steve Jobs and later by Disney. Since the beginning, the company's identity has been an intense dialogue between art and technology. At first the technological side was dominant. In its early years, Pixar was mostly a computer hardware company. Its Pixar Image Computer was designed to perform graphic design for hospitals and medical research facilities, but at $135,000 it was too expensive to become successful. In a remarkable example of innovative cross-fertilization, an employee named John Lasseter began producing computer-animated shorts, seeking to demonstrate the visual power of the technology. In 1984 Lasseter showed a short film called *The Adventures of André and Wally B.* at an industry convention. It was a sensation. Everyone immediately recognized the movie as a major leap forward for the computer animation industry.

Pixar had found its true vocation. It shed the hardware side and embraced moviemaking. Today Pixar's campus, squeezed between pharmaceutical labs and biotech startups, is a factory of imagination and a theme park. It boasts the most commercially successful blend of innovators, artists, and geeks that has ever existed in the film industry. John Lasseter is now the company's chief creative officer and has directed many of its biggest box office successes. The media love him, both for his artistic genius and for his colorful Hawaiian shirts. He is clearly not one of the suits.

While Pixar no longer makes computer hardware, the creative tension between art and technology persists and is probably at the root of the company's success. The making of each movie is a constant back-and-forth between the project's artistic and technical sides. The job of the technical people like Glynn is to protect the artistic side by developing techniques that allow the story, the characters, and the visuals to blossom. This is innovation at its best, a fusion of technical creativity and artistic expression that generates something new and valuable. Through technological and creative advances, Pixar has revolutionized the art of animation. In the process, it has become a household name, achieved unprecedented commercial success, and garnered almost universal praise from movie critics around the world. It has received twenty-six Oscars, more than two for each movie produced.

Ultimately, Pixar's success has to do with the talent and creativity of people like Glynn. In an era when much of what we consume is produced in Asia, the manufacturing of dreams still takes place in California. With its Disneyland-like atmosphere, its corporate garden dominated by giant *Toy Story* puppets, and the anarchic atmosphere of the animators' dens, Pixar feels quintessentially American. It is hard to imagine company headquarters moving to Shenzhen.

What Is an Innovation Job?

Over the past fifty years, the U.S. economy has gradually shifted away from traditional manufacturing toward the creation of knowledge, ideas, and innovation. As traditional manufacturing jobs keep disappearing, the innovation sector keeps growing. It will soon be what manufacturing used to be in the 1950s and 1960s: America's main engine of prosperity.

During the 1980s and the early 1990s, global innovation was generally stable, with the worldwide number of patents at around 400,000 per year. But since 1991 global investment in research and development has been increasing. The number of patents granted around the world exceeded 800,000 in 2010 and reaches new heights almost every year. In the United States, the top patent producers in 2010 were IBM (5,866 patents), Microsoft (3,086), Intel (1,652), and Hewlett-Packard (1,480). The largest category of patents was pharmaceutical (highlighting the importance of the life science industry for American innovation), followed by information technology, chemical and material sciences, scientific instruments, telecommunications, and, much lower in the ranking, newer industries such as nanotechnology. To find categories associated with traditional manufacturing, you have to dive deep, to positions 37 and 38: land vehicles and metalworking. Compared with the 2010 ranking, the 1992 ranking looks prehistoric. In 1992, IBM was already near the top, but IT and life sciences were much less prominent, while companies making traditional manufacturing products and photographic equipment were among the top innovators. Canon, Fuji, and Kodak were all among the top ten patent producers that year.

Jobs in the innovation sector are not easily defined, because innovation takes many forms. Of course they include the high-tech sector: information technology, life sciences, clean tech, new materials, robotics, and nanotechnology. But jobs in innovation also include parts of the labor market outside of science and engineering. As with Dominic Glynn and Pixar, jobs in innovation are often found in unexpected places. What they all have in common is that they make intensive use of human capital and human ingenuity.

A growing number of skeptics have questioned the importance of innovation for the American economy, arguing that the increase in jobs is not large enough to offset the losses in manufacturing. Intel's former CEO Andy Grove has famously criticized America's "misplaced faith in the power of startups to create U.S. jobs." Tyler Cowen's influential book *The Great Stagnation* argued that companies like Facebook or Twitter do not have many employees, because they rely on their users for most of the content and are simply too small to replace the titans of the past, like Ford and General Motors.

But the picture that emerges from the data is more complex. Take employment in the Internet sector. Before even looking at the numbers, I suspected that Internet jobs had to be growing. After all, the Web has become our favorite place to get news, buy products, search for information, connect with each other, and look for partners. However, I was not prepared for the dramatic nature of the employment growth. Using data from a comprehensive data set of all U.S. businesses collected by the Census Bureau, I estimate that the number of jobs in the Internet sector has grown by 634 percent over the past decade, or more than two hundred times the growth rate of the overall number of jobs in the rest of the economy during the same period. (This number does not even include Internet-related jobs outside the high-tech sector, such as the delivery of online purchases.) As you can see in Figure 3, this rate of job growth is explosive and has been accelerating over the past five years. If the rest of the labor market had grown like the Internet sector, not only would there be no unemployment, there would be two new job openings for each citizen, including babies and the elderly. The growth of total salary earned in this sector has been even more dramatic—712 percent over a ten-year period in today's dollars.

Skeptics are correct in pointing out that Facebook directly employs only 1,500 workers in its Menlo Park headquarters and another 1,000 scattered around America. While this number is growing rapidly, General Electric and General Motors have about 140,000 and 79,000 employees in the United States, respectively. But Facebook is just a platform, and most of the apps that make it attractive are created by other companies. Some, such as the game-maker Zynga, have more employees than Facebook. A recent study

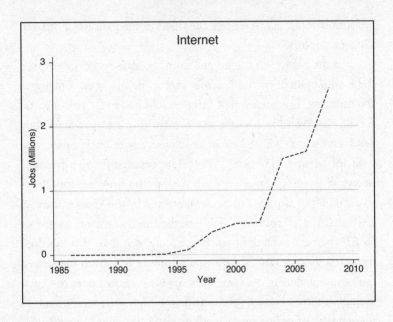

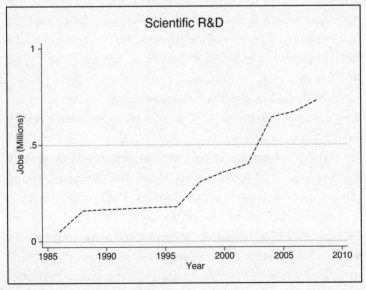

Figure 3. The rise of jobs in innovation

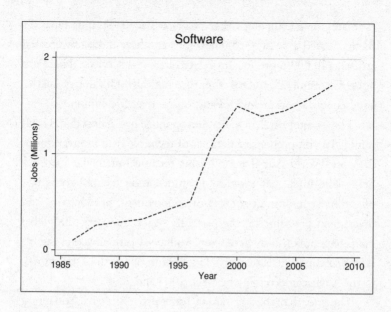

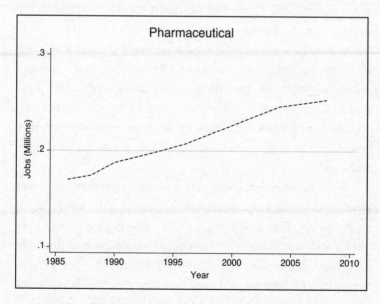

Figure 3 (cont.). The rise of jobs in innovation

estimates that companies that produce Facebook apps have directly created at least 53,000 new jobs and have indirectly created at least 130,000 more jobs in related business services. These are not trivial numbers, and together they put the total value of salaries and benefits associated with Facebook at over $12 billion.

The economist Michelle Alexopoulos has painstakingly assembled a comprehensive data set of technological innovations in the post–World War II period using technical manuals in various fields. She finds that progress in information technology is one of the most important sources of employment, productivity, and investment growth over the past fifty years. Similarly, the global management consulting company McKinsey estimates that the Internet sector alone is responsible for about one-fifth of the growth of the American economy between 2004 and 2008.

The growth of the software sector is also impressive. You might not realize it, because most of the media coverage focuses on the outsourcing of software jobs to places like Bangalore. But the data tell us that U.S. jobs in software have actually grown by 562 percent over the past two decades—not as explosive as the Internet sector, but still thirty-three times greater than the rest of the labor market. (The important exception is programming. Programmers, who are typically less skilled than software engineers and computer scientists, have been much more exposed to "offshoring" and automation.)

With an impressive 300 percent growth in employment over twenty years, life science research is another pillar of the innovation sector. This figure only includes jobs in private-sector research and development—biotech, for example—and does not take into account researchers who work at universities or government labs. The Bureau of Labor Statistics puts biomedical engineers at the top of the list of the twenty occupations expected to grow the most over the next ten years, with a predicted growth rate of 72 percent. Medical scientists, biochemists, and biophysicists rank near the top.

Advanced manufacturing—anything from robotics and pharmaceuticals to electronics and advanced medical devices—constitutes another important part of the innovation sector. Companies such as Apple, IBM, and Cisco are, after all, manufacturers, and a large part of all private R&D in America comes from advanced manufacturing. A recent study shows that advanced manufacturing companies make up a small fraction of all manufacturing companies but are the ones that add the most value and have the highest productivity. Job growth is slower but more stable than in younger parts of the innovation sector. For example, the fourth panel in Figure 3 shows the steady gains enjoyed by pharmaceutical manufacturers over the past three decades. While not all parts of this sector are creating jobs—as we saw with computers and semiconductors—advanced manufacturing is doing better than traditional manufacturing. Crucially, the mix of jobs is changing rapidly, with fewer positions for blue-collar workers and more for engineers, designers, and marketers. This reflects the diffusion of iPhone-style supply chains in which goods designed in America by American companies and built using American technologies are physically produced abroad.

But the innovation sector encompasses more than science and engineering. It includes parts of industries as diverse as entertainment, industrial design, marketing, and even finance. In the past three years alone, venture capital has invested about $2 billion in two hundred financial service startups. That amounts to "tens of thousands of jobs being created," according to the CEO of Bill-Float, one of the industry leaders. Take Prosper, a peer-to-peer lending startup that connects individual borrowers and individual lenders, thus creating loans with more advantageous terms than those available from mainstream banks. On a random day not long ago, Prosper's listings included a California mother seeking $5,000 to go back to school, an Arizona artist seeking $4,000 to buy a truck to haul paintings to his shows, a Napa Valley winemaker seeking $4,000 to buy new oak barrels for the 2011 harvest, and a Michigan TV producer seeking funds to improve his reality TV show. From a technological point of view, Prosper's innovation is not particularly sophisticated: it is just a website where those who need money meet those who have money. But from a social point of view, it is revolutionizing the way small businesses and families can get credit. In the process, it is creating jobs both at Prosper itself and at the businesses it supports.

In 2010, 3,649 patents were granted in the United States for financial or business practice innovations, one of the most important categories. Currently the American public holds a rather negative view of the social value of financial "innovation," given the role of derivatives in triggering the Great Recession of 2008–2010. But this is probably an overreaction. While there are important exceptions, by and large financial innovation has contributed to supporting America's economic growth. After all, the fact that planes are flying people from one corner of the country to another at affordable prices has as much to do with innovative hedges against high fuel costs as with advances in aviation technology.

Digital entertainment is another fast-growing piece of the innovation sector. More than eight hundred people work at Pixar alongside Dominic Glynn. And although not everyone can work in the world's most cutting-edge film studio, the digital entertainment industry as a whole has created thousands of good jobs over the past twenty years. When *Star Wars* was filmed in 1976, special effects consisted of plastic model starships set against hand-painted backgrounds of galaxies. Starting in the 1990s, special effects moved decisively to the digital realm. Today a growing number of movies, TV shows, and ads include some digital component. The digital music business is also growing fast and includes established companies such as Pandora as well as a wealth of small startups that are reshaping the way we experience music, from playlist systems based on social networks to innovative file-sharing systems and from mobile karaoke to new tools to play rock star in your garage. And then, of course, there are video games. When I was a kid, video games involved crude green-on-black graphics and repetitive scratching noises. Today they have become visually striking reproductions of reality. With revenues exceeding $20 billion a year—more than the film and music industries combined—games are big business and account for tens of thousands of jobs.

In the end, it does not matter whether American workers make something physical, like more efficient lithium batteries for electric cars, or something immaterial, like a better search engine. What really matters is that American workers produce goods or services that are innovative and unique and not easily reproduced. This is the only way to generate jobs that pay well in the face of stiff global competition.

Why Innovation Matters to You

I am claiming that innovation has become America's new engine of prosperity. But what does that really mean? What exactly is an "economic engine"? It is important to clarify that an economic engine is not necessarily the largest sector of an economy. Estimates of the total number of innovation jobs differ, depending on how exactly *innovation* is defined, but a reasonable estimate is that about 10 percent of all jobs in the United States belong to the innovation sector. While that number is growing, the innovation sector will never constitute the majority of our employment. Put simply, the average American worker will never be employed by an Internet startup or Pixar. Even manufacturing at its peak never employed more than 30 percent of the U.S. labor force.

The reason is simple: the vast majority of jobs in a modern society are in local services. People who work as waiters, plumbers, nurses, teachers, real estate agents, hairdressers, and personal trainers offer services that are produced and consumed locally. This sector exists only to serve the needs of a region's residents and is largely insulated from national and international competition. Economists call this the *non-traded sector*. Such jobs are "non-tradable" because they cannot be exported outside the region where they are produced: you need to consume them where you produce them.*

Take yoga. Today yoga is big business, and it is growing. Jennifer Aniston recently declared to *People* magazine, "Yoga completely changed my life," and she is not alone. Many stars, including Madonna and Sting, are true believers, together with an estimated 15.8 million people who practice yoga regularly, up from 4 million only ten years ago. This industry is generating yearly revenues of about $6 billion in classes, retreats, private instruction, and even yoga cruises. As the writer Mary Billard put it, "Zen is expensive." From the point of view of yoga purists, this may sound sacrilegious. But from the point of view of job creation, it's gold. Tens of thousands of people work in the United States as yoga teach-

---

* Here I am defining the non-traded sector from a local perspective. The definition of *non-traded sector* that is more commonly used by economists is the national one. Not all jobs that are traded at the city level are also traded at the national level. For example, national news providers such as MSNBC, Fox News, and NPR offer a service that is largely traded from the point of view of a city but non-traded from the point of view of the nation. The definition is not always perfect. Most restaurant jobs are in the non-tradable sector, since they cater to local residents. But in tourist destinations such as Las Vegas, restaurant jobs are in the traded sector, as they mainly serve outsiders.

ers, making up part of the 261,000 Americans considered "fitness workers" today. This number is expected to grow rapidly in the foreseeable future as Americans make more and more use of yoga centers, health clubs, and fitness facilities.

Yoga instructors are a small part of a vast web of non-tradable jobs. In the United States, two-thirds of all jobs are in this sector. Most of the 27 million jobs created over the past two decades have been in the non-tradable sector, with health care as the fastest-growing. Even in Silicon Valley, residents are more likely to work in a store than for a high-tech firm.

By contrast, most jobs in innovative industries belong to the traded sector, together with jobs in traditional manufacturing, some services—parts of finance, advertising, publishing—and agricultural and extractive industries such as oil, gas, and timber. These jobs, which account for about a third of all jobs, are very different, because they produce a good or service that is mostly sold outside the region and therefore needs to be competitive in the national and global marketplace. For example, Microsoft and Boeing export most of their products to customers who do not reside in Seattle. Google, too, provides a service—Web search—that is mainly used outside its headquarters in Mountain View, California.

The paradox is that while the vast majority of jobs are in the non-traded sector, this sector is not the driver of our prosperity. Instead, our prosperity mainly depends on the traded sector. There are two reasons for this. The first is that productivity growth is different in the two sectors. As I mentioned earlier, in many parts of the non-traded sector, labor productivity does not grow very much. The number of yoga instructors needed to teach a class today is the same as it was fifty years ago and will probably never change. A therapy session today takes as long as it did in the time of Freud; the amount of labor needed to paint a house, fix a leaking pipe, babysit a child, or sell real estate is more or less the same as it has always been. Although parts of the non-traded sector experience productivity increases (improvements in medical technology, for example, have made doctors and nurses more productive), the more typical case is one of limited productivity increases. By contrast, productivity in the traded sector tends to increase over time, thanks to technological progress. As we have seen, it takes 75 percent fewer worker hours today than it did in 1950 to make a car. Labor productivity in the high-tech sector grows even faster, thanks to a constant stream of innovation.

The debate on jobs often misses this key point. This productivity difference between traded- and non-traded-sector jobs matters because, as we saw, the only way to raise workers' standard of living is to raise their productivity. Interestingly, higher productivity of workers in the traded sector means higher salaries not just for the workers in that sector but also for workers in other sectors, especially those with similar skills. Historically, when manufacturing wages inched up, other sectors had to adjust to remain competitive. For example, builders needed to raise the wages of carpenters, roofers, and plumbers to keep them from taking a manufacturing job, even though productivity in construction was flat. So even if the manufacturing sector accounted for a minority of the workforce, for decades it was an engine strong enough to lift the salaries of many American workers, including those who worked in services. From this perspective, it becomes clear why its demise is so terrifying. And it becomes equally clear why the rise of innovation is so crucial. It is more than just the jobs in that sector that are at stake—it's the entire economy.

There is a second, related reason that the rise of innovation matters to all of us. While the first reason reflects forces that are national in scope, this second reason reflects forces that are local but equally important. Every time a company generates jobs in the innovation sector, it also indirectly creates additional jobs in the non-traded sector in the same city. Attracting a new scientist, software engineer, or mathematician to a city increases the demand for local services. This in turn means more jobs for cabdrivers, housekeepers, carpenters, nannies, hairstylists, doctors, lawyers, dog walkers, and therapists. These local service workers cluster around high-tech workers, supporting their personal needs. In essence, from the point of view of a city, an innovation job is more than a job.

To see how this multiplier effect works in practice, let me introduce you to a small-business owner named Tim James. James is a bookbinder in San Francisco. His clients are mostly local residents and local businesses, so he is clearly part of the non-traded sector. He employs eight workers who bind books and do custom printing. His employees are good with their hands and tend to have low levels of education. If you visit his cavernous, neon-lit shop, the first things you notice are several beautiful old-style cutting and binding machines that dominate the floor. Paper is everywhere, with some pieces stacked in neat piles on the ground, ready to be used, and others in small strips. Dust covers the machines and the floor. Bookbinding appears to be a very labor-intensive craft. The technology used in James's shop has not changed much in the past thirty years.

Although James's shop is definitely low-tech, the performance of his business over the years closely tracks the ups and downs of the NASDAQ, which in turn closely tracks the performance of high-tech firms in San Francisco. James's business soared during the late 1990s, the dot-com boom years. During that period, high-tech workers flush with cash filled local restaurants and bars, built new houses, and gathered in local gyms, thus vastly increasing the incomes of local service workers, including bookbinders. To keep pace with the rising demand for his products, James hired three new employees and raised everyone's wages. During the dot-com bust that followed, demand for James's products—and therefore the number of his employees—dropped, only to recover more recently with the expansion of the local high-tech sector.

James's experience is not unique. Indeed, it perfectly exemplifies the strong link between innovation jobs and local services. With only a fraction of the jobs, the innovation sector generates a disproportionate number of additional local jobs and therefore profoundly shapes the local economy. A healthy traded sector benefits the local economy directly, as it generates well-paid jobs, and indirectly, as it creates additional jobs in the non-traded sector. What is truly remarkable is that this indirect effect on the local economy is much larger than the direct effect. My research, based on an analysis of 11 million American workers in 320 metropolitan areas, shows that for each new high-tech job in a metropolitan area, five additional local jobs are created outside of high tech in the long run.

I mentioned this earlier, but it gets even more interesting. These five jobs benefit a diverse set of workers. Two of the jobs created by the multiplier effect are professional jobs—doctors and lawyers—while the other three benefit workers in nonprofessional occupations—waiters and store clerks. Take Apple, for example. It employs 12,000 workers in Cupertino. Through the multiplier effect, however, the company generates more than 60,000 additional service jobs in the entire metropolitan area, of which 36,000 are unskilled and 24,000 are skilled. Incredibly, this means that the main effect of Apple on the region's employment is on jobs *outside* of high tech. (Incidentally, Apple is among Tim James's clients: when Steve Jobs died, James was commissioned to make the family's condolence book.) In essence, in Silicon Valley, high-tech jobs are the *cause* of local prosperity, and the doctors, lawyers, roofers, and yoga teachers are the *effect*. It is pretty simple: at the end of the day, someone has to pay for all those yoga sessions.

All parts of the traded sector have a multiplier effect, but innovation has the largest. My analysis indicates that attracting one job in traditional manufacturing generates 1.6 additional local service jobs—less than a third of the corresponding figure for high tech. Ron Bloom, President Obama's former manufacturing czar, liked to say, "If you get an auto assembly plant, Walmart follows; if you get a Walmart, an auto assembly plant does not follow." He is correct: the manufacturing sector does generate local service jobs too, and this is a major benefit for communities. But he misses the fact that if a community were to attract an Internet or a biotech company of similar size, the effect on job creation in the service sector would be even larger. Not only would it create three times as many jobs, but those would be better-paying than Walmart jobs. Take a city like Seattle. Although a manufacturing company such as Boeing has twice as many jobs in Seattle as Microsoft does, it ultimately creates fewer local jobs.

How can the high-tech multiplier effect be so much larger than that of other industries? What is so special about high tech? To begin with, high-tech workers are very well paid, with salaries and benefits typically considerably above the average. This means they consume more local services than other workers and therefore create more local jobs. With more disposable income, these employees go to restaurants, visit hairdressers, and see therapists more often. According to a company report, the annual compensation of the average employee at Microsoft is $170,000. This is an incredibly high figure, especially if you consider that it takes into account everyone in the company, including secretaries and janitors. After subtracting what an employee spends on nonlocal goods, housing, taxes, and savings, this leaves about $80,000 available to be spent on local services. This amount alone can support two local nonprofessional jobs at prevailing wages. In addition to employees' personal consumption, high-tech companies' operations require many local business services, and this means more graphic designers, marketers, business consultants, and security guards.

The final reason for the large high-tech multiplier effect is that high-tech firms tend to be located near each other. Bringing one high-tech company to a city eventually results in having more high-tech companies locate there, as dense high-tech clusters make high-tech firms more innovative and more successful. This clustering effect also exists in manufacturing, but it is particularly strong in high tech, for reasons that we will discover soon. The end result is the creation of more local service jobs and an even larger multiplier effect.

Policymakers and business leaders love to praise the virtues of American small businesses, invariably pointing out that small businesses are responsible for most job creation. While this is true, the vast majority of small businesses are in retail and other non-traded services. In the end their existence is dependent on the vitality of the traded sector, where large businesses are dominant. There would not be many retail jobs in a city if not for the income generated in the traded sector.

The multiplier effect is a remarkable feature of the labor market. The current public debate about the American economy is often framed in terms of an inherent tension between the interests of one group and the interests of another: the two Americas of the rich and the poor, the haves and the have-nots. While this tension may be true in the case of fiscal policy—for example, when we are deciding on how much to tax high-income earners—in most other cases it is a false juxtaposition. As far as job creation is concerned, there is no inherent contradiction between the interests of high-income workers and those of low-income workers. Indeed, the key lesson of the multiplier effect is that the economy is a tightly interconnected system, and what is good for one group typically tends to be good for another. This is a case where the rising tide does lift all boats—at least those boats that are in the same city.

New Jobs, Old Jobs, and Recycled Jobs

One of the key reasons the innovation sector—unlike manufacturing—has so many jobs is that even today it remains a remarkably labor-intensive sector. The main production input in scientific research is human capital—in other words, people and their ideas. Writing software still requires hours of typing on a keyboard. All you have to do is visit a workplace to see which production inputs are really important. In a factory, the dominant components of the factory floor are clearly the machines, and everything else, including the location and activities of the people who operate them, revolves around them. In a lab or a software company, it's clearly the people who matter, and everything revolves around *them*. Ironically, the workplaces where the most innovative technologies are created are still largely dependent on human labor, while the workplaces where traditional goods are made are largely run by robots.

Take digital entertainment, for example. Adding digital effects to a movie requires hours and hours of creative labor. I realized just how labor-intensive the process was when I went to see *Avatar* with a programmer named Kent Matheson, who had worked on the film at George Lucas's special effects company, Industrial Light and Magic. Matheson had spent weeks designing one of the starships that appeared on the screen for just a few seconds of the two-and-a-half-hour-long movie. My good friend Ben Von Zastrow works a few miles north of Pixar in a smaller independent studio called Tippett Studio. His job is to create computer-generated small furry animals for movies. Technically, he is a lighting artist. Of course, there are no physical lamps or sets; Ben uses software to shed light on animated creatures. He works alongside a team of other digital artists with evocative names like animator, modeler, texture painter, compositor, and puppeteer, but all of them create images by typing on a keypad. It took Ben and his team three full months to create something as simple as a small wolf for the movie *Twilight: New Moon*.

The fact that creating images with a keyboard remains so laborious is not a bad thing for the workers in this sector. On the contrary, it is a very good thing, because it means more jobs for them—at least for now. Of course, the day will arrive when some new software will enable digital artists to create starships and furry animals in weeks instead of months, and then in days, and eventually in hours or minutes. In the beginning, digital artists will be thrilled, because they will be able to do their jobs much more easily and quickly. But in the long run it will mean fewer jobs. In the same way that tractors and combines replaced farm workers and robots replaced factory workers, powerful computers and better software will someday do the work of digital artists.

Jobs that today look more like art than labor will eventually become commodified, standardized, and mechanized—they will lose their luster, and their numbers will start to shrink. This is likely to happen not just in digital entertainment but in most other parts of the innovation sector, as technological progress reduces the need for human labor. We must hope that when this happens, promising new ideas and products will appear and the cycle of old and new jobs will start all over again.

A common fallacy is to think that all jobs in innovation, because they involve new technologies, are "new jobs." But in many cases they are just replacing existing jobs, and in some cases they might even result in fewer jobs overall. For example, travel websites have undeniably created social value, because they have made booking plane tickets and hotels cheaper and more convenient. Thousands of "new" Internet jobs have been created for those who design and manage the sites of companies such as Expedia and Travelocity, but the job losses have been far greater, as countless travel agencies across the country have had to shut down. Similarly, Netflix has increased the variety of rental movies available, but this is proving disastrous for thousands of neighborhood video stores.

Innovation keeps churning out an ever-changing roster of jobs, and yet the net effect is positive. An analysis of the French Internet sector found that since the advent of the Web, the Internet has created 1.2 million jobs (both for positions directly linked to the Internet, such as software engineering, and positions outside the sector, like delivery of online purchases) and destroyed 500,000, thus generating a net gain of 700,000. In other industrialized countries, the best estimates are that 2.6 jobs are typically created for every one destroyed. Importantly, although the job losses are geographically widespread, the job gains are mostly concentrated. In the case of travel websites and Netflix, Seattle, New York, and the San Francisco Bay Area have experienced employment gains, since this is where the websites tend to be located, while all other cities have suffered a loss of retail jobs.

Why Jobs in Innovation Will Keep Increasing

In 2007 a twenty-seven-year-old entrepreneur named Sam Lessin cofounded the Web startup drop.io, which was intended to make real-time file sharing and collaboration easy for people. Three years later, in a major coup, Lessin sold his startup to Facebook. Immediately after paying millions of dollars for the company, Facebook did something unexpected: it shut down drop.io. As it turns out, what Facebook had wanted all along was Sam Lessin. It is part of a new phenomenon that has emerged in Silicon Valley: large established companies buy entire startups not to acquire groundbreaking new technologies but to acquire the people who thought of them in the first place.

This is typically great news for those working in the companies that are acquired, as it translates into generous salaries and stock options. The *New York Times* recently reported that in 2009, Facebook bought FriendFeed, a company that helps people track the online activities of their friends. "Tech insiders thought it was trying to compete more effectively with Twitter. But Facebook was really after FriendFeed's dozen well-regarded product managers and engineers," including its cofounder Bret Taylor. The price tag for FriendFeed is estimated to be $47 million, or $4 million per employee. "We really wanted to get Bret," Mark Zuckerberg, Facebook's chief executive, remarked at the time. "Someone who is exceptional in their role is not just a little better than someone who is pretty good," he went on to say. "They are 100 times better."

Zuckerberg's comments are particularly revealing. The rise of the innovation sector is associated with an increase in the value of talent, for a simple reason: economic value depends on talent as never before. In the twentieth century, competition was about accumulating physical capital. Today it is about attracting the best human capital. What Zuckerberg was effectively saying is that the economic returns on new ideas have never been so high, and the rewards for those who can come up with good ideas have also increased. "Engineers are worth half a million to one million," Vaughan Smith was reported as saying in the same *New York Times* article. He should know. As Facebook's director of corporate development, Smith was behind Facebook's acquisitions of more than twenty "talents" in the past four years.

Why have the economic returns on new ideas increased so much? After all, you would think that even thirty years ago having a new idea would have created a great deal of economic value. What has changed? Fundamentally, there are two reasons for this increase: globalization and technological progress. Remarkably, the same two forces that have caused the demise of blue-collar jobs are now fueling the rise of jobs in the innovation sector.

Increased globalization is particularly good news for innovative companies. The reason is simple, although often ignored in the political debate on jobs and globalization. Innovative industries are fundamentally different from all other industries in how they make their profits. Take software: coming up with an idea for a new piece of software, developing it, and testing it is expensive, but when the software is written, it can be reproduced millions of times at virtually no cost. Most costs incurred by Microsoft in developing new versions of Windows—the backbone of its commercial success—involve paying engineers to write software codes. These costs are largely fixed, in that they do not depend on how many copies of Windows are sold. The variable costs—the cost of the physical CD-ROM on which Windows is actually written and the cost of the cardboard box in which it is shipped—are trivial. This means that it costs Microsoft billions of dollars to make the first copy but only a few cents to make the second. A global market enables the company to sell vastly more copies without increasing production costs. The same is true for Internet services, pharmaceuticals, digital media, and most products that involve significant up-front R&D costs. Google spends millions every month to improve its search engine. The cost of this investment does not vary whether ten people use the company's website or one billion use it. The only thing that varies is its profits. Similarly, most of the cost of making a new drug entails up-front research investment. The cost of making the actual pill is minor.

In most innovative industries, the main production costs are the fixed costs of research and development. The variable costs of production are typically low. Having access to global markets dramatically raises the returns on creating new ideas by increasing sales without increasing costs. It is therefore not surprising that the resources devoted to innovative activities have reached unprecedented levels. This is important not just for the bottom line of innovative companies but for the creation of jobs. These features of innovative industries stand in contrast to traditional manufacturing, where fixed costs can be large but variable costs are also significant. For example, when making cars or clothes, each additional unit significantly adds to total costs. Thus the benefits of larger markets are less pronounced in traditional manufacturing than in innovation.

The effect of globalization is strengthened by the expansion of the global middle class. As countries such as China, Brazil, and India become more prosperous, they demand more high-end products. This trend tends to favor innovative industries. American exports to China have increased almost 500 percent, more than ten times faster than exports to the rest of the world. A large percentage comes from California, Washington, and Texas in the form of advanced products such as software, scientific instruments, medical machines, and aerospace products.

Still, numbers aside, there is widespread unease about globalization. Surveys consistently indicate that the majority of Americans, including those in the innovation sector, believe that globalization is one of the main causes of America's economic problems. Eric Scott is a good example. He is an experienced hardware engineer and has worked in high tech most of his life. A couple of years ago he landed a good position at Dolby Labs, a high-tech company in San Francisco that makes digital sound systems for cinemas and audio applications for DVDs. Scott has a wife, a three-year-old daughter, and a new mortgage on a house not too far from the Dolby R&D facility. He has noticed that over the years Dolby has been experimenting with outsourcing parts of the innovation process to cheaper locations in Asia. So far it does not appear that this outsourcing process has taken off in his company, but he can easily imagine a future in which most innovative work, including his own, is outsourced. Is this where we are heading?

Until recently, most developing countries have focused on labor-intensive, low-skill manufacturing sectors that compete primarily on price. But at some point they may get tired of being mere producers of goods "designed in California." China already produces more patents than Germany and France. To be sure, the quality of these patents, as measured by citations by other patents, is still low compared to those of Western countries.* Nonetheless, it is undeniable that innovation in China and India is steadily increasing over time.

While outsourcing causes job losses in most parts of traditional manufacturing, the opposite is true for the innovation sector. How is this possible? Recent research shows that more assembly jobs in China and more customer assistance jobs in India ultimately mean more R&D jobs in America as well as more jobs for the professionals—advertisers, designers, analysts, accountants—who cluster around high-tech companies. Let's consider Oracle, the giant maker of business hardware and software. In 2000, Oracle had 22,008 workers in the United States and 20,919 workers abroad. Today Oracle has 40,000 workers in the United States and 66,000 abroad. While the U.S. share has declined, the actual number of American jobs has increased. Crucially, the best-paid jobs, the ones in R&D, are still overwhelmingly in the United States, and

---

* Part of the reason is that the Chinese government has created financial incentives to file for patents, irrespective of the quality. *The Economist* reports that tenure decisions for academics take into account the number of patents filed, and patent office employees receive bonuses if they approve many patents. Corporate income taxes are lower, and the probability of winning government contracts is higher for firms that file many patents. It is perhaps not too surprising that the firm that files the largest number of patents in the world is the Chinese company Huawei.

their number has increased significantly. As James Fallows once put it, Indian and Chinese workers "making $1,000 a year have been helping American designers, marketers, engineers, and retailers making $1,000 a week (and up) earn even more. Plus, they have helped shareholders of U.S.-based companies."

To see how this works in practice, consider the case of life sciences research. Over the past decade, many American biotech and pharmaceutical companies moved part of their R&D activity abroad. Dr. Vinita Sharma is the head of the Indian government's National GLP Compliance Monitoring Authority, Department of Science and Technology, which is India's equivalent of the U.S. Food and Drug Administration. She is an intense, cosmopolitan woman with vast experience in research and policy. She has a vision: Indian-made R&D services on demand. When I met her, she was in an excellent mood. India was about to receive certification from the OECD to operate preclinical test labs for pharmaceuticals, industrial chemicals and agrochemicals. The certification meant that India could now offer a vast array of laboratory tests to European and U.S. life science companies. "You have an idea in the U.S., Germany, or China, we will provide you with good R&D," she said. "We want to convert knowledge to wealth."

This is exactly the sort of initiative that terrifies American workers. How will it affect their jobs? The effect could be positive or negative, depending on whether the hiring of more foreign R&D workers induces American companies to hire more or fewer workers in the United States. In economics jargon, it depends on whether foreign workers are complements or substitutes for workers here. A series of studies by the Dartmouth economist Matthew Slaughter suggest that outsourcing is not a zero-sum proposition, because overseas workers are generally a complement to rather than a substitute for U.S. workers: for every job outsourced by American multinationals, nearly two new jobs are created in the

United States. This is a good deal for American workers. Those new jobs tend to be positions in research and development, marketing, engineering, design, and science. They command good salaries and offer significant potential for career advancement. One of the most comprehensive reports on offshoring in technical fields, produced by the National Academy of Engineering, agrees: "Offshoring appears to have contributed to the competitive advantage of U.S.-based firms in a variety of industries."

Globalization is not the only reason that jobs and salaries in innovation are growing. When a company successfully brings an innovation to the market—anything from the iPad to a new drug—it can often charge a price that is significantly higher than production costs. Economists call this an *economic rent*. In this respect, an innovative product is like a Versace handbag. In high fashion, the rent comes from the glamour of the brand. In high tech, the rent comes in the form of a patent that gives the innovator monopoly power.

Who ultimately benefits from the economic value created by innovations? Consumers benefit in the form of new or cheaper products. Companies benefit in the form of higher profits. The rest accrues to the workers involved in developing the product. This means more jobs and in some cases higher salaries. The economists Natarajan Balasubramanian and Jagadeesh Sivadasan used a highly confidential data set compiled by the Census Bureau to follow the inner workings of 48,000 American companies over twenty years. They found that both employment and labor productivity grow significantly in the year after a firm successfully patents its first innovation and that these positive effects persist for years afterward. The London School of Economics professor John Van Reenen examined the relationship between salaries and innovation in six hundred innovative firms in the United Kingdom. Focusing on innovations that are both technologically important and commercially successful, he found that the average salary in a company increases substantially as a result of innovation, peaking about three years after the introduction of a new product.

Thus the economic rent created by innovations ends up benefiting not just CEOs and shareholders but also workers. The salary gains are substantial: overall, Van Reenen estimates that workers capture about 20 to 30 percent of the additional economic value created by innovation in the form of higher salary. This is one of the reasons that innovation is so important as an engine of job creation. As we saw in the case of the iPhone, there is little value in making standard products that can be produced anywhere in the world. But when the traded sector in a country makes products that are innovative and unique, it creates more and better jobs.

The supply of skilled and creative workers capable of innovating is increasing worldwide, as a growing number of young people in emerging economies obtain college and postgraduate education. But the demand for skilled and creative workers is rising even faster. The latest recession temporarily slowed this increase in demand, but in the long run, globalization and technological progress mean more jobs and greater rewards for the creative workers who produce new ideas and new products. While this is good news for American society as a whole, the effects of this shift on American workers are geographically uneven. The creation of new jobs is not spread uniformly over the entire country. It favors some cities and regions while ignoring others. Geography is becoming increasingly important. In the next chapter we will look at who wins and who loses in the new innovation economy and how these trends are reshaping the social fabric of American communities.

3

◆

THE GREAT DIVERGENCE

Seattle is not the same city that it was thirty years ago. Dilapidated warehouses have been rehabilitated to host scores of small startups, and smart-looking new office buildings have been erected for larger companies. The once crumbling pier and rusting docks now house Internet and software companies. The old rail yard has been renovated and is now home to the labs of the pharmaceutical company Amgen. A formerly seedy residential and commercial district on the northern edge of downtown has become the cool city's new zip code, with well-designed offices and condominiums appearing every year.

If you sit for a couple of hours outside the Victrola Coffee and Art café on 15th Avenue in the Capitol Hill district, you will get a good sense of Seattle's unpretentious energy and discreet optimism. With a diverse mix of thirty-something professionals, gay couples, struggling college students, and wealthier, more established families, the area is one of the city's many thriving neighborhoods. People stroll along lively streets dotted with eclectic book-stores and bodegas specializing in artisanal goods. Throughout the city, gourmet restaurants and new cultural venues have taken over abandoned structures and surface parking lots. Even Pioneer Square, until recently known more for its methadone clinics than for trendy startups, is experiencing a renaissance as high-tech companies such as Zynga, Discovery Bay, and Blue Nile move into its beautiful historic brick buildings. It is becoming so fashionable that it is even attracting financial institutions: Maveron, the venture capital firm of Starbucks chairman Howard Schultz, has just taken over a renovated space on First Avenue.

Seattle is a place that mixes a strong sense of community with a contagious entrepreneurial energy and an understated cosmopolitan vibe. Above all, it exudes a quiet confidence about the future, a confidence rooted in one simple fact: Seattle has completely transformed itself from a decaying old-economy provincial town into one of the world's preeminent innovation hubs. In the process, its residents have become some of the most creative and best-paid workers in the United States.

This has not always been the case. It is hard to imagine today, but a visitor walking Seattle's streets three decades ago would have had a completely different impression of the city. In the late 1970s, Seattle was looking inward and backward, consumed by fear about the future, riddled with crime, and decimated by job losses. But on a rainy morning in early January 1979, something happened that changed the history of the city.

A Tale of Two Cities

Everyone now associates Microsoft with Seattle. However, in the early part of its life, Microsoft was located a world away. In fact, the company was founded in 1975 in Albuquerque, New Mexico. That year it had one product, one client, and three employees. The client was MITS, an Albuquerque hardware firm that was making a successful home computer kit called the Altair 8800; the product used BASIC software to operate the kit. In the following months and years, Microsoft prospered in New Mexico. Its future looked so promising that by the end of 1975 one of the two founders, an intense, preppy-looking twenty-year-old named Bill Gates, took a leave of absence from Harvard to join Paul Allen, the other founder, who was already in Albuquerque. The business took off, and Gates never returned to Harvard to finish his degree. Not that he needed it. Sales were growing exponentially, and by 1978 they had already exceeded the $1 million mark and thirteen employees worked for Microsoft.

But the founders were growing increasingly restless, and eventually they decided to relocate. This was not a business decision. Gates and Allen were both from Seattle, and they both wanted to go back to the place where they had grown up. On New Year's Day 1979, the company packed its bags and moved to Bellevue, a sleepy suburb on the other side of Lake Washington from Seattle.

In 1979, Seattle was not an obvious choice for a software company. In fact, it seemed like a terrible place. Far from being the high-flying hub it is today, it was a struggling town. Like many other cities of the Pacific Northwest, it was bleeding jobs every year. It had high unemployment and no clear prospects for future growth. It was closer to today's Detroit than to Silicon Valley.

Just like Detroit, Seattle's problem was simple: its economy was heavily dependent on old-style manufacturing and lumber, a decidedly unattractive industry mix. Just like Detroit, about half of its manufacturing jobs were in transportation. Unsurprisingly, its employers were having a hard time and were downsizing. People were leaving the city by the thousands. Historically, aerospace had always had a strong presence in Seattle, anchored by Boeing and several subcontractors, but during the 1970s and early 1980s, Boeing suffered multiple slumps. Paccar, a large truck manufacturer and another major local employer, was also experiencing problems. There were some bright spots, such as Nordstrom headquarters and the port, but their presence was simply too weak to lift the local economy. With the exception of people employed by Boeing and the University of Washington, Seattle residents were not particularly skilled.

As a result, the quality of life was declining. Today most people think of Seattle as one of the most pleasant cities in America, weather aside. But when Microsoft moved there in the late 1970s, the crime rate was significantly higher than in Albuquerque and there were 50 percent more robberies per capita. The quality of Seattle schools was mixed, museums were run-down, and the culinary scene, now so interesting and eclectic, was unremarkable. Starbucks, at the time a tiny local company with only three stores, was still serving the standard watered-down American brew and had yet to ignite the espresso revolution.

Just a few years earlier, *The Economist* had labeled Seattle the "city of despair." In an article on the alarming decline of the local economy, its correspondent reported that "the country's best buys in used cars, in secondhand television sets, in houses, are to be found in Seattle, Washington. The city has become a vast pawnshop, with families selling anything they can do without to get money to buy food and pay the rent." Expectations about the future were so low that a giant billboard appeared near the airport saying, "Will the last person leaving SEATTLE—Turn out the lights." The billboard, which is still very much talked about today, perfectly captured the mood of a city in decline.

Although the relocation of Microsoft from Albuquerque to Seattle seemed insignificant at the time, it helped turn Seattle into one of America's most successful innovation hubs. What is remarkable is how serendipitous it was. Bill Gates and Paul Allen could have moved the company to Silicon Valley, where many other technology companies were already established, or they could have stayed in Albuquerque. With its dry weather, relaxed attitudes, the Sandia National Laboratories, and the University of New Mexico, Albuquerque seemed as if it was destined to develop a local high-tech cluster, and it probably would have if Microsoft had stayed there. From the point of view of Microsoft, staying in Albuquerque would not have been crazy in 1979. The idea of a move met some resistance at first, because some of the employees liked New Mexico and did not want to deal with the logistics, but Gates and Allen stood firm in their decision.

More than any other sector, innovation has the power to re-shape the economic fates of entire communities, as well as their cultures, urban form, local amenities, and political attitudes. We know this, and yet determining the precise interplay of all these forces and distinguishing cause and effect is difficult, especially in complicated places like Silicon Valley. By contrast, the history of the high-tech sector in Seattle can be traced to one specific and fortuitous event, which makes it an insightful natural experiment.

Before the move, the labor markets in Seattle and Albuquerque looked quite similar. In 1970, for example, the number of college-educated workers in Seattle was only 5 percent higher than in Albuquerque, relative to population. Salaries were also slightly higher in Seattle, because of all the Boeing engineers and the large number of hospitals and clinics associated with the University of Washington, but the differences were small and the trends were similar in the two cities. After the move, the paths of these cities started diverging in irreversible ways. By 1990 the difference in the number of workers with a college education had grown to 14 percent, and in 2000 to 35 percent with the explosion of the high-tech sector. It has now reached a staggering 45 percent. This is an enormous difference, similar to the one that exists between the United States and Greece. Importantly, salary levels have also been diverging, especially among skilled workers. In 1980 college graduates in Seattle were making just $4,200 more than college graduates in Albuquerque; they are now making $14,000 more.

Since Albuquerque lost Microsoft, its economy has limped along. Modest gains in the schooling of its workforce have hampered the growth of the local innovation sector. Intel and Honeywell have large production facilities there and Bank of America and Wells Fargo have large back offices in town, but far more typical are low-end jobs in low-value-added services. By and large, Albuquerque's innovation cluster never reached the critical mass needed to sustain a truly competitive high-tech ecosystem. By contrast, Seattle has one of the largest concentrations of software engineers in the world. This agglomeration is so large that more than one-quarter of the salaries paid to North America's software workers are in Seattle. T-Mobile, the fourth-largest wireless carrier in the United States, has a presence both in the Seattle area and in Albuquerque. But Seattle has the company headquarters—with all the high-paying jobs and their large multiplier effect—while Albuquerque has a customer service center, with many low-end jobs and a small multiplier effect.

As the economic fortunes of the two cities continue to diverge, all other aspects of daily life—from livability to cultural amenities, from school quality to food quality—are also growing apart. Although Albuquerque was safer than Seattle in 1979, its violent crime rate is now higher than Seattle's, and its murder rate is more than double.

What happened to these two cities exemplifies the diverging economic paths experienced by many American cities over the past three decades. Because of the self-sustaining nature of economic development, cities that are similar initially can become very different over time as small differences become magnified. Winners tend to become stronger and stronger, as innovative firms and innovative workers keep clustering there, while losers tend to lose further ground. Economists have a term for this: *multiple equilibria*.

How exactly does it happen? This is the part of the story that is the most important. Microsoft employs 40,311 people in the Seattle area, 28,000 of whom are engineers engaged in R&D. This may sound impressive, but how can 40,311 jobs possibly change the destiny of a metropolitan area of almost 2 million residents? The answer is that Microsoft's ultimate effect on the local economy is much larger than the number of people it employs. First, when Microsoft moved to Seattle, the city increased its attractiveness to other high-tech companies. Microsoft effectively serves as the anchor of the local high-tech sector and a magnet for other software companies. The history of Amazon is interesting in this respect. In 1994, Amazon's founder, Jeff Bezos, lived in New York and was vice president of a large and successful Wall Street firm. Although he had a job that paid what most people only dream of, he felt there was even more he could achieve. It was the beginning of the Internet era, and he wanted a piece of the action. Bezos eventually quit his job and started an Internet book retailer. He decided to name it after the earth's longest river, the Amazon, and to locate it in Seattle.

Why Seattle? When Bill Gates made his choice, Seattle was an unattractive place to start a high-tech company, but he had personal reasons to be there. By contrast, Jeff Bezos had no personal reason to be in Seattle. He was not born there (in fact, he was born in Albuquerque!). But by the time he started his company, fifteen years after Gates's move, Seattle had become a magnet for high-tech activity. Because Microsoft was in the city, software engineers and programmers had concentrated there in large numbers and venture capital firms had opened offices there. At a time when people who knew how to create good websites were still rare, Bezos found real talent in Seattle. He also found financing. The first non-family investor in Amazon was a Seattle-based venture capitalist named Nick Hanauer, whose $40,000 funding played a crucial role in helping the company survive its delicate early phase. Shortly after, another Seattle-based venture capitalist provided $100,000 to make the new website more user-friendly, which gave the nascent company a key competitive advantage.*

Microsoft did not directly help Jeff Bezos start his company, but its presence triggered the creation of an entire high-tech cluster in the region. This highlights a remarkable feature of the high-tech world: success generates more success. It is a feature that has enormous implications for the future of many cities, and it is the main theme of this chapter and the next one. The moment Bezos left Manhattan and headed west, a series of events began that would ultimately bring thousands of good jobs to Seattle. Today the little outfit that Bezos started in his garage is a global brand with 51,000 employees worldwide, a third of whom are in Seattle.

A second channel through which Microsoft reshaped the local economy was by spawning a host of other companies in the Seattle area, as millionaire employees quit to launch their own firms. By one estimate, Microsoft alumni alone have started four thousand new businesses, the majority of them in the Puget Sound area. Expedia is one example of a local company that spun directly out of

---

* Seattle also offered a tax advantage. Internet retailers did not have to charge sale taxes to consumers outside their state, which made locations just outside large states with many consumers—like California—particularly competitive. Still, this advantage could not be the deciding factor, since dozens of cities closer to California than Seattle is—Las Vegas, Phoenix, Boise, Portland, Eugene—enjoy the same tax advantage but never developed a significant Internet sector.

Microsoft. RealNetworks is another. Founded by former Microsoft employee Rob Glaser in 1995, it now has 1,500 employees and is one of the largest private employers in the city. In his spare time, Jeff Bezos founded a human space-flight company called Blue Origin. Located just twenty minutes outside Seattle, it is like something out of a movie: a private company that builds and flies spacecraft.

But what if you are not a rocket scientist, software engineer, or computer scientist? What does all this mean for the average worker in Seattle? Because of the multiplier effect, Microsoft's most notable impact on the Seattle labor market has been on workers employed outside the high-tech sector. I estimate that Microsoft is responsible for creating 120,000 jobs for service workers with limited education (cleaners, taxi drivers, real estate agents, carpenters, small-business owners) and 80,000 jobs for workers with college or advanced degrees (teachers, nurses, doctors, and architects). These numbers have been increasing over time because Microsoft salaries keep rising and because the company's demand for local services keeps growing.

Innovation creates enormous social benefits, in the form of new drugs, better ways to communicate and share information, and a cleaner environment. These benefits are diffuse, in the sense that consumers all over the world can enjoy them. But innovation also creates benefits in the form of new and better jobs. These benefits are overwhelmingly concentrated in a small number of geographic locations. Of course, not all of these changes are for the better, and later we will look more closely at housing costs and gentrification. But first we need a clearer picture of the geography of innovation jobs in the United States. Seattle is certainly not the only innovation hub in America. To get a handle on where the jobs of the future will be, we need to figure out where innovation is taking place now.

Where Are the Hubs?

One hundred years ago the hot new technology was the automobile, a seemingly miraculous new machine that promised to change the world. Initially thousands of small producers were scattered across the country. A few decades later the number dropped to three giant corporations, with most of the production near Detroit. Today car factories are again spread all over the world, from Brazil to Poland. When personal computers first appeared in the 1970s, a myriad of small independent producers were scattered all over America. Steve Jobs and Steve Wozniak made the first Apple computer in 1976 by buying components from a mail-order catalogue and assembling them in Jobs's garage. Later the production of personal computers became a highly concentrated industry, with just a few key players, mostly in Silicon Valley. Right now the industry is maturing, and production is scattered among hundreds of low-cost locations. The same pattern has been documented in industries as diverse as iron founding, flour milling, and cigarette production.

Just like people, industries have life cycles. When they are infants, they tend to be dispersed among many small producers spread all over the map. During their formative years, when they are young and at the peak of their innovative potential, they tend to concentrate to harness the power of clusters. When they are old and their products become mature, they tend to disperse again and locate where costs are low. Thus it is not surprising that the innovation sector—the part of the economy that is now going through its formative years—is concentrated in a handful of cities.

One way to map innovation today is to look for the inventors. Every time an inventor files a new patent, he is required to report his residence. These data on patents are publicly available and provide some interesting statistics. Of course, not all new ideas are patented and not all patents are great innovations, but economists have long used the concentration of patents as a proxy for the creation of new products and ideas.* The accompanying map (overleaf) shows the metropolitan areas that have been generating the most innovation relative to their size, as measured by the number of patents per resident.

The first things you will notice are the large differences between different parts of the country. The map shows clusters of intense innovative activity (the dark areas) surrounded by an ocean with almost no innovation activity (the light gray and white areas). The states that generate the most patents are California, New York, Texas, and Washington, with California producing the lion's share and New York a distant second. These four states alone generate almost half of all patents granted in the United States, up from a third in 1980.

Among metropolitan areas in the United States with at least a million residents, the most innovative one by far is the San Francisco–San Jose region, which includes Silicon Valley. Its lead over the runner-up, Austin, Texas, is enormous. The average resident in the San Francisco–San Jose region produces more than twice as many patents as the average resident of Austin and more than three hundred times the number of patents as the average resident in McAllen, Texas, the city with the lowest number of patents per

---

* Patents are an imperfect proxy for two reasons. First, many innovations, especially those outside science and engineering, are never patented. Second, many patents never turn into commercially valuable innovations. To account for the latter, economists sometimes use patent citations to assign more weight to influential patents. The idea is that if a particular patent is cited by several other patents, it is likely to be more important than a patent that nobody cites. Using weights does not change the information here very much, but it does significantly alter international comparisons, as we saw earlier with China.

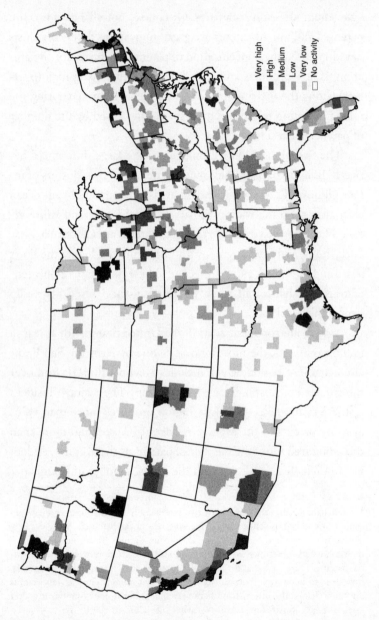

Map 1. Patents per capita, by metropolitan area

capita. This difference between the most and least innovative metropolitan areas is truly staggering. New Orleans is near the bottom of the list, together with the Norfolk–Virginia Beach–Newport News area, Miami, Las Vegas, and Nashville. These differences are not limited to patents but extend to other measures of innovation, including venture capital and jobs. The San Francisco–San Jose region has more than four times as many high-tech jobs as Austin. Compared with cities at the other end of the spectrum, the divide in high-tech employment is immense.

In the aftermath of the dot-com bust of 2001–2003, pessimism about the future of America's high-tech clusters abounded. Observers predicted the end of the Valley's global dominance. But the pessimists were by and large wrong. Silicon Valley has remained the innovation capital of the world, and it continues to lead all other metropolitan regions in the breadth and scope of its innovative activity. It accounts for more than a third of all venture capital investment, significantly more than twenty years ago. Every year hundreds of smart, ambitious innovators move their startups from Europe, Israel, and Asia to Silicon Valley. The Valley keeps its position as the world's number-one innovation hub not because those who are born there are smarter than anyone else but because of its unparalleled power to attract great ideas and great talent from elsewhere.

After Silicon Valley, Austin stands out. It is a latecomer, but its rate of growth, spurred by computer and electronic products, has been spectacular over the past two decades. Dell is one of the key employers, and many other global high-tech companies have offices there, including IBM, 3M, Applied Materials, Advanced Micro Devices, and Freescale Semiconductor. Austin appears to complement rather than compete with Silicon Valley. The two are linked by a constant flow of college-educated high-tech professionals who shuttle between the two regions. Although they are not particularly close geographically, San Francisco–San Jose is the most popular destination for Austin residents with a college degree who relocate.

Two other top performers are Raleigh-Durham and Boston-Cambridge. Anchored by excellent research universities and world-class medical facilities, they boast impressive concentrations of scientific R&D services and life science innovation. Their success is driven by the ability of local entrepreneurs to turn the academic life sciences research done at Harvard, MIT, and Tufts in the Boston area and at Duke and the University of North Carolina in the Raleigh–Durham–Chapel Hill area into commercial ventures. In addition, Boston has maintained a strong presence in the design of precision instruments and has developed a software cluster. When Microsoft opened its first East Coast research lab in Cambridge in 2008, the company remarked that the main reason was the "large community of scientists in New England, notably the faculty and students at the many premier academic institutions in the vicinity."

San Diego is an interesting example of a city that has evolved over the past thirty years from a small community of retirees and surfers to one of the world's most geographically concentrated biotech clusters, revolving around the Scripps Research Institute, the Salk Institute, and the University of California at San Diego. It includes biotech giants such as Amylin Pharmaceuticals as well as dozens of midsized biotech firms with promising portfolios of new drugs. It also includes a number of jobs in electronic hardware for telecommunications. Although the Middlesex-Somerset-Hunterdon area in New Jersey also generates a large number of patents in the life sciences, it has a completely different feel from San Diego. With an unsurpassed concentration of established pharmaceutical and medical companies, including Bristol-Myers Squibb and Johnson & Johnson, its corporate landscape is dominated more by powerful incumbents than by up-and-coming startups.

In this respect, New York City and Washington, D.C., are anomalies. Like all real estate in New York, lab space is scarce, which limits the amount of scientific R&D performed in the city. Thus New York is not among the metropolitan areas with a large number of patents per capita. But with about 300,000 technical jobs, New York remains one of the key world-class innovation hubs and is getting stronger over time. During the past two decades, its Silicon Alley has become a magnet for creative entrepreneurs and well-educated young workers. It is now a premier location for Internet portals and information services, and along with Los Angeles it accounts for a large number of digital entertainment jobs. In 2011, Google undertook a major expansion of its New York office, paying $2 billion to buy a large building near the meatpacking district. The New York region also remains the undisputed world leader in financial innovation.

Historically, the Washington, D.C., area never had many high-tech jobs apart from a sizable cluster of defense contractors. But in the past twenty years the region has been remarkably successful at attracting a wide array of innovative companies to the high-tech Dulles corridor and to the downtown area. It is not on the list of the top ten metropolitan areas for patents because its industry mix is focused on IT and generates relatively fewer patents than other places. But there are almost 300,000 high-tech workers in D.C., double the average concentration in the United States. Life science companies increasingly flock to the area, attracted by proximity to high-profile public institutions such as the National Institutes of Health.

Although not many people realize it, Dallas has been rising in high-tech rankings because of a major concentration of telecommunications jobs, a solid semiconductor presence anchored by Texas Instruments, and a growing data-processing cluster. Smaller but emerging high-tech clusters can be found in Minneapolis, Denver, Atlanta, and Boise. While few regions excel in many areas of high technology, America is dotted by clusters that specialize in one or two. For example, Rochester, New York, home to Kodak and Xerox, focuses on optical technology, while Rochester, Minnesota, home to the Mayo Clinic, focuses on medical research. Dayton, Ohio, has become a center for radio frequency identification; Salt Lake City, Bloomington, and Orange County specialize in medical devices, Albany in nanotech, Portland in semiconductors and wafers, and Richmond, Kansas City, and Provo in information technology.

America's innovation hubs are a very diverse group. On the surface, it is not immediately obvious what they might have in common. The lifestyle in San Diego is quite unlike the one in New York City or Boston. Salt Lake City and San Francisco have very different cultures and almost opposite political values. Seattle and Dallas share little in terms of amenities. But if we dig a little deeper, it becomes clear that all these cities have one thing in common: they have a very skilled labor force and therefore a remarkably productive traded sector. As we are about to discover, this means more and better-paying jobs for everyone who lives there.

Your Salary Depends More on Where You Live Than on Your Résumé

Here is a question for you: Which large city pays computer scientists the most? As you might expect, computer scientists in San Jose and San Francisco, the country's high-tech capitals, are the best paid in the nation (and the world). The average computer scientist in San Francisco and San Jose makes $130,000 annually. The same person in Boston or New York or Washington, D.C., would make 25 to 40 percent less.

Here is another question: Which city pays lawyers the most? I must admit that I got this one wrong. Before looking at the data, I was expecting New York and Washington, D.C., to have the highest-paid lawyers in America. The image I had was of high-powered lawyers in $5,000 tailor-made suits brokering billion-dollar deals in the centers of finance and power. And although New York and Washington do have the largest number of lawyers among American cities, they do not have the highest-paid lawyers. Using data collected by the Census Bureau, I discovered that San Jose is the city where lawyers earn the most—an average of over $200,000 a year—and San Francisco is not too far down the list. Lawyers in cities at the other end of the spectrum—Albany, Buffalo, and Sacramento—tend to make less than half what lawyers earn in San Jose.

For waiters, the place to be is Las Vegas. A waiter in one of the city's most luxurious restaurants can earn six figures. But even waiters working in normal establishments do well in Sin City. The typical waiter makes $18.20 an hour, tips included—the highest average hourly wage in any large metropolitan area. This is probably not too surprising: in one of the world's preeminent adult entertainment destinations, Las Vegas waiters benefit from the generous tipping associated with gambling and other vices. The cities that come next on the list are more telling. San Francisco, Seattle, Boston, and Washington, D.C., are ranked numbers two through five. San Diego is number seven. Of the top ten cities for waiters, three are purely tourist destinations (Las Vegas, Orlando, and West Palm Beach) but seven are cities with a strong high-tech presence.

Remarkably, the same is true of other jobs, in both the traded and non-traded sectors. The ranking for industrial production managers is dominated by San Jose, Austin, Portland, San Francisco, Raleigh-Durham, and Seattle—all innovation hubs. For barbers and hairstylists, San Francisco, Boston, and Washington, D.C., are in the top five. These workers earn on average 40 percent more than their counterparts in Riverside and Detroit. For cooks, Boston tops the list, with an average yearly salary of $31,782, and Houston and San Antonio are at the bottom, at about $20,000. For architects, San Francisco leads the pack. If you detect a pattern here, you are right.

In principle, the lawyers, hairstylists, and managers in Boston and San Francisco could just be better at their jobs than the ones in Houston, Riverside, and Detroit. Maybe they are more experienced, smarter, or more motivated. But the salary differences do not change all that much when we take into consideration work experience, level of education, or even IQ. The workers themselves aren't that different; what is different is the local economy that surrounds them—especially the number of skilled workers.

Forty years ago, the rich areas in America were manufacturing capitals with an abundance of physical capital. Cleveland, Flint, and Detroit had significantly higher average incomes than Raleigh and Austin. Today human capital is the best predictor of high salaries both for individuals and for communities. Average income in Raleigh-Durham and Austin is significantly higher than in Cleveland, Flint, and Detroit. The presence of many college-educated residents changes the local economy in profound ways, affecting both the kinds of jobs available to residents and the productivity of all workers. In the end, this results in high wages not just for the skilled workers but also for workers with limited skills. This is the most surprising part of the story, and it deserves to be told in some detail.

The workforces in American communities have vastly different skill levels. Map 2 shows just how different. It displays the percent

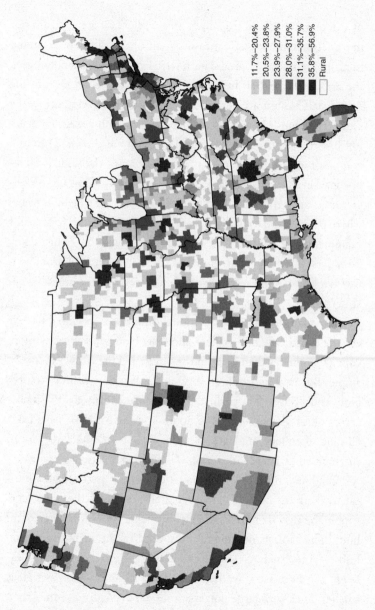

11.7%–20.4%
20.5%–23.8%
23.9%–27.9%
28.0%–31.0%
31.1%–35.7%
35.8%–56.9%
Rural

Map 2. Share of workers with a college degree, by metropolitan area

of workers in each metropolitan area with a college education or more. To obtain a precise measurement of the differences among metropolitan areas, I used data on 15.4 million workers between the ages of 25 and 60 living in 306 metropolitan areas from the American Community Survey, which is collected every year by the Census Bureau. The census defines metropolitan areas to include not just the political boundaries of a city but also its neighboring communities, to the extent that they are part of the same local labor market as suggested by commuting patterns. Thus metropolitan areas are economically integrated regions that include places where people tend to live and work. The New York metro area, for example, includes New York City and its suburbs in Long Island, New Jersey, Connecticut, and Westchester County. (Throughout this book, I use the terms *city* and *metropolitan area* interchangeably.)

The map shows marked regional differences in schooling levels: the Northeast and coastal California tend to have many more college graduates than cities in the South and the Midwest, for example. But even more interesting is the fact that within each region, and even within each state, there are enormous differences across cities. For example, the South and the Midwest have large clusters of college graduates in cities such as Atlanta and Denver, surrounded by areas with very few college graduates.

Imagine ranking American cities by the percentage of local workers with a college degree. Table 1 shows the large metropolitan areas that are at the top of such a list—America's brain hubs. The table is dominated by several of the main innovation hubs—cities such as Washington, D.C., Boston, San Jose, Raleigh, San Francisco, Seattle, Austin, and Minneapolis—as well as smaller cities with large universities, such as Madison, Ann Arbor, Fort Collins–Loveland, and Lincoln. In these cities, almost half of the labor force is college-educated, and a significant fraction has a postgraduate degree. Portland, New York, and Denver are also in the top group.

Table 2 shows the metropolitan areas that are at the bottom of the list. This group includes Vineland-Milville-Bridgetown, New Jersey; Yuma, Arizona; Flint, Michigan; and of course Visalia, the California community that we encountered at the beginning of this book. In these cities only one in ten workers has a college degree and there is virtually no high-tech presence.

The sheer size of the differences between American communities is staggering. Stamford, Connecticut, the city with the largest percentage of college-educated workers in the United States, has *five times* the number of college graduates per capita as the city at the bottom, Merced, California. This difference is enormous and much larger than the difference that we typically see in European countries. Indeed, it's much larger than the difference in schooling levels between the United States and many developing countries, such as Sri Lanka (three times), Bolivia (three times), and Ghana (four times). It is not just the American-born residents that are creating this education gap. The majority of immigrants who settle in the first group tend to be highly educated professionals, while the majority of immigrants in the second group have low levels of schooling.

These facts are not just fodder for dinner-party conversations. Differences in educational levels are associated with huge differences in salary. The tables indicate that college graduates in brain hubs make between $70,000 and $80,000 a year, or about 50 percent more than college graduates in the bottom group. Compare San Jose, number five from the top, with Merced, at the very bottom. Both cities are in California, less than 100 miles apart, but their labor markets belong to two different universes. San Jose, in the heart of Silicon Valley, has more than four times the number of college graduates per capita as Merced and salaries that are 40

TABLE 1: METROPOLITAN AREAS WITH THE LARGEST
SHARE OF WORKERS WITH A COLLEGE DEGREE

	Rank	Percentage with college degree	Salary of college graduates	Salary of high school graduates
Stamford, CT	1	56%	$133,479	$107,301
Washington, DC/MD/VA	2	49%	$80,872	$67,140
Boston, MA/NH	3	47%	$75,173	$62,423
Madison, WI	4	47%	$61,888	$52,542
San Jose, CA	5	47%	$87,033	$68,009
Ann Arbor, MI	6	46%	$65,452	$55,456
Raleigh-Durham, NC	7	44%	$63,745	$50,853
San Francisco–Oakland, CA	8	44%	$77,381	$60,546
Fort Collins–Loveland, CO	9	44%	$57,391	$47,007
Seattle-Everett, WA	10	42%	$68,025	$55,001
Trenton, NJ	11	42%	$81,914	$64,299
Lexington-Fayette, KY	12	41%	$55,238	$44,915
Austin, TX	13	41%	$62,289	$48,809
Portland, OR	14	40%	$57,366	$48,080
Minneapolis–St. Paul, MN	15	40%	$69,955	$57,187
Denver-Boulder, CO	16	39%	$64,488	$50,097
New York–Northeastern NJ	17	38%	$79,757	$59,797
Lincoln, NE	18	38%	$50,401	$41,837
Santa Cruz, CA	19	38%	$64,801	$48,186
Tallahassee, FL	20	38%	$59,380	$46,715
Worcester, MA	21	37%	$60,723	$48,465

TABLE 2: METROPOLITAN AREAS WITH THE SMALLEST SHARE OF WORKERS WITH A COLLEGE DEGREE

	Rank	Percentage with college degree	Salary of college graduates	Salary of high school graduates
Mansfield, OH	286	17%	$53,047	$35,815
Beaumont–Port Arthur–Orange, TX	287	17%	$58,234	$38,352
Rocky Mount, NC	288	16%	$52,330	$34,329
Stockton, CA	289	16%	$59,651	$37,928
Fort Smith, AR/OK	290	16%	$50,937	$33,187
Ocala, FL	291	16%	$47,361	$32,725
Yuba City, CA	292	16%	$56,403	$34,999
Modesto, CA	293	15%	$60,563	$36,126
Waterbury, CT	294	15%	$54,651	$37,280
Brownsville–Harlingen–San Benito, TX	295	15%	$43,800	$22,450
McAllen-Edinburg-Pharr-Mission, TX	296	15%	$44,605	$22,845
Anniston, AL	297	15%	$48,928	$33,031
Yakima, WA	298	15%	$50,160	$29,084
Bakersfield, CA	299	14%	$65,775	$34,807
Danville, VA	300	14%	$42,665	$28,868
Houma-Thibodaux, LA	301	14%	$56,044	$37,395
Vineland–Milville-Bridgetown, NJ	302	13%	$57,668	$35,375
Flint, MI	303	12%	$43,866	$28,797
Visalia-Tulare-Porterville, CA	304	12%	$55,848	$29,335
Yuma, AZ	305	11%	$52,800	$28,049
Merced, CA	306	11%	$62,411	$29,451

percent higher for college graduates and a whopping 130 percent higher for workers with a high school diploma.

Or compare Boston, third from the top, with Flint, fourth from the bottom. Both have a proud industrial past, but their economies are now at opposite ends of the spectrum. Boston, with four times the number of college graduates, is heavily dependent on innovation and finance. Flint, with one of the smallest concentrations of human capital in the nation, is still focused on traditional manufacturing, primarily cars. A college graduate in Boston makes on average $75,173, or 75 percent more than the salary of a similar worker in Flint. Of course, the relationship between innovation and salaries is not perfect. Stamford's wealth is mostly due to financial services, while salaries in Raleigh, one of the world's top innovation hubs, are relatively low. Nevertheless, there is a clear tendency for cities with many college-educated residents to have a local economy with a great deal of innovation and good salaries.

Possibly the most remarkable fact shown by these tables is that high school graduates in the top group often make more than *college* graduates in the bottom group. The average worker with a high school education living in Boston makes $62,423, or 44 percent more than a college graduate in Flint. A high school graduate in San Jose earns $68,009, thousands of dollars more than college-educated workers in Merced, Yuma, Danville, and all the other cities at the bottom. In other words, the disparity between cities is so large that it can dominate the disparity between levels of education. This underscores the fact that wage differences in the United States have as much to do with geography as they have to do with social class.

Today, America's economic map shows not one but three Americas. At one end of the spectrum are the brain hubs in Table 1, with high salaries for both skilled and unskilled workers. At the other extreme are communities in Table 2, with low skill levels and declining labor markets. In the middle, a number of cities appear undecided about which direction to take, as their future could go either way. Note that it is not just that brain hubs pay high average salaries because they have many college-educated residents and these residents earn high salaries. There is something more at play here. Brain hubs pay high average salaries to *nonskilled* workers too. Thus, a worker's education has an effect not just on her own salary but also on the entire community around her.

How can this be possible? One answer is that the cost of living in cities like San Jose, Raleigh-Durham, and Austin is higher than in Flint or Merced, and high school graduates need to be compensated to live there. This is true—higher cost of living does offset some of these differences—but it is not the whole story. It explains why there are still people in Flint and Merced and why not everyone has moved to San Jose, Raleigh-Durham, and Austin. But it does not explain why there are still employers in San Jose, Raleigh-Durham, and Austin. Why should employers, especially those who compete nationally, put up with such high labor costs to be in these locations? We will come back to the issue of cost of living and what it means for people's standard of living. But first we need to dig a little deeper into the relationship between a city's educational level and its economic prospects. What explains it, and what does it mean for communities?

How Your Neighbor's Education Affects Your Salary

The link between local human capital and salaries is robust, and it holds true for most American cities. Figure 4 shows the relationship between average salaries of high school graduates in each city and the fraction of workers with a college education in that city. The graph shows a clear positive association, indicating that the more college graduates there are, the higher the salaries for high school graduates are. (The outlier in the top right corner is Stamford. Because there are 305 other cities in the graph, the relationship is not driven by this outlier.) The economic effect is quite large. The earnings of a worker with a high school education rise by about 7 percent as the share of college graduates in his city increases by 10 percent. For example, a worker with a high school education who moves from a city like Miami, Santa Barbara, or Salt Lake City, where 30 percent of the population are college graduates, to a city like Denver or Lincoln, where 40 percent of residents are college graduates, can expect a raise of $8,250 just for moving.

When I first looked at the graph, I was concerned that it was an apples-to-oranges comparison — that workers who pick cities with many college graduates, like Boston, might be fundamentally different from workers who pick cities with fewer college graduates, like Flint. If Boston attracts high school graduates who are

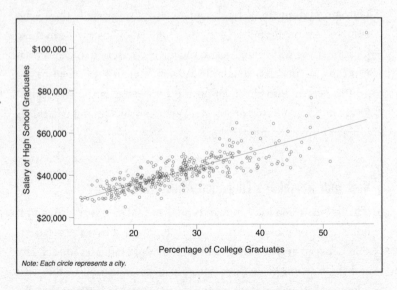

Figure 4. The relationship between salaries of high school graduates and the percentage of college graduates in a city

smarter or more ambitious than those in Flint, then we should not be surprised to find out that they earn more. To account for this possibility, I relied on fourteen years of data from the National Longitudinal Survey of Youth, which has followed the life histories of 12,000 individuals since 1979. This data set is particularly useful, because it ensures an apples-to-apples comparison by tracking how the salary of a given person changes over time as the number of college graduates in his city changes. I found that workers who live in cities where the number of college graduates increases experience faster salary gains than workers who live in cities where the number of college graduates stagnates. Thus the same individual can make a very different salary depending on how many skilled workers surround him. This relationship holds for all sectors, but it is particularly strong for workers with high-tech jobs.

This is a truly remarkable finding, one that helps explain the vast differences in the economic success of various cities. There are three reasons for the relationship between the number of skilled workers in a city and the wages of their unskilled neighbors. First, skilled and unskilled workers complement each other: an increase in the former raises the productivity of the latter. In the same way that working with better machines increases a worker's productivity, working with better-educated colleagues increases the productivity of an unskilled worker. Second, a better-educated labor force facilitates the adoption of newer and better technologies by local employers. Third, an increase in the overall level of human capital in a city generates what economists call *human capital externalities*.

This concept is at the heart of modern economic growth theory, the study of what determines a country's economic success. Researchers have built sophisticated mathematical models showing that sharing knowledge and skills through formal and informal interaction generates significant knowledge spillovers. These knowledge spillovers are thought to be an important engine of

economic growth for cities and nations. In a famous 1988 article, the Nobel laureate Robert Lucas argued that these spillovers
may be large enough to explain long-run differences between rich
and poor countries. His explanation was that when people interact,
they learn from each other, and this process makes those who interact with better-educated peers ultimately more productive and
creative. This human capital externality is a financial windfall that
people collect simply because they're surrounded by many educated people.

The sum of these three effects—complementarity, better
technology, and externalities—is what ultimately drives the positive relationship in Figure 4. Notably, this relationship is strongest for less skilled individuals. In a study that I published in 2004,
I found that for a college graduate, an increase in the number of
other college graduates in the same city does result in a salary increase, but not a particularly large one. For a high school graduate, the increase is four times larger. For a high school dropout, the
effect is five times larger. Thus, the lower the skill level, the larger
the salary gains from other people's education.

A large number of highly educated workers in a city is also associated with more creativity and a better ability to invent new
ways of working. One way to see this is to look at what Jane Jacobs called "new work," novel occupations that did not exist before. The economist Jeffrey Lin has studied which cities are the
most creative in America, in the sense that they generate the most
"new work" as measured by jobs that did not exist ten years earlier.
Examples of new work in 2000 include Web administrator, chat-
room host, information systems security officer, IT manager, biomedical engineer, and dosimetrist (don't ask: I have no idea what
a dosimetrist does). Between 5 and 8 percent of workers are engaged in new work at any time, but this number is much higher
in cities that have a high density of college graduates—the ones

in Table 1—and a diverse set of industries. Lin also found that creativity pays off: for the first few years after a new kind of job is created, workers in those positions earn significantly higher wages than identical workers in old jobs.

The existence of human capital externalities is good news for less educated workers in highly educated cities, because it means that they end up earning more than they would otherwise. But it also implies that well-educated individuals are not fully compensated for the social benefits that their education generates. This is an important example of a market failure. Essentially, education has a private benefit, in the form of higher earnings for the individual who acquires it, and an additional benefit for all other individuals who live in the same city. In fact, the full return on education for society—sometimes called *social return*—is larger than its private return. Since college graduates are not compensated for the benefit that they bestow on everyone around them, there are fewer college graduates than we as a society would ideally like. To put it differently, if the salary of college graduates reflected its full social value, more people would go to college. One way to correct for this market failure is to provide public subsidies for college education. Indeed, this is the reason that state and local governments pick up much of the cost of educating their residents. There are certainly other reasons to justify public investment in higher education—political and ethical—but I know of none more powerful than this one. It is in our own interest to subsidize other people's education, as it ends up indirectly benefiting us.

The Great Divergence and the New Geography of Inequality

What I find most striking about the socioeconomic differences between the three Americas is that they are not going away. Instead, the divergence among our communities is deepening and accelerating. Cities and states that start off strong tend to become relatively stronger, and cities that start off weak tend to become relatively weaker.

Map 3 shows the change in the percentage of workers with a college education since 1980 for each metropolitan area in the country. Boston was already well educated in 1980; since then it has increased its share of college-educated workers by 23 percentage points, a jump of more than two-thirds relative to its 1980 level. Stamford has outperformed all other cities, doubling its percentage of college-educated workers. By contrast, since 1980, Visalia and Merced have added only one percentage point to their shares of college-educated workers. It is hard to believe, but Flint has not added college-educated workers to its labor force in thirty years. While the rest of the country has become better and better educated, Visalia, Merced, and Flint have been treading water—and barely staying afloat.

This Great Divergence is among the most significant developments in recent American economic history. As communities grow apart, the U.S. population is becoming more and more segregated, not across urban neighborhoods but across cities and regions. With every passing year, college graduates are increasingly settling in cities where many other college graduates already reside, while high school graduates are increasingly settling in cities where many other high school graduates reside. A good way to appreciate the Great Divergence visually is Figure 5, which shows the gains since 1980 in the percentage of college-educated workers for the ten most educated cities and the ten least educated cities in each year. In the past three decades, the top group has experienced large gains, while the bottom group has grown much less.

Evidence indicates that American cities are more racially integrated today than at any time in the past century, a trend that has been accelerating in the past two decades. In 2010, for example,

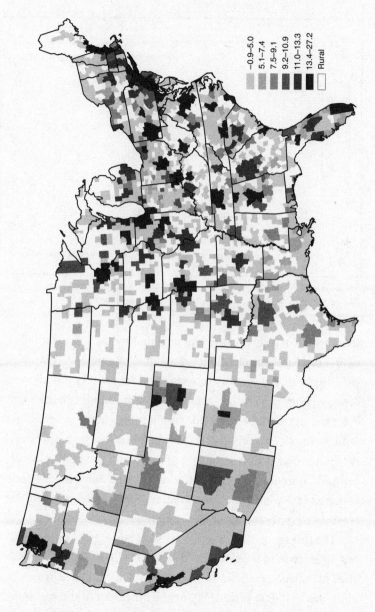

-0.9–5.0
5.1–7.4
7.5–9.1
9.2–10.9
11.0–13.3
13.4–27.2
Rural

Map 3. *Changes in the share of workers with a college degree, 1980–2008, by metropolitan area.* Note: *These changes are measured in percentage points.*

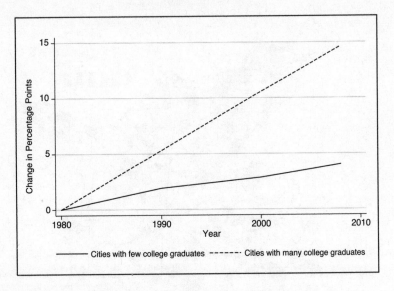

Figure 5. Gains in the share of college graduates since 1980

there were virtually no all-white neighborhoods, and the number of predominantly black neighborhoods has plummeted. It is somewhat ironic that at the very moment that our neighborhoods were desegregating racially, our country was segregating educationally. This has tremendous economic implications, but also social and political ones. A country that is made up of regions that differ drastically from one another will end up culturally and politically balkanized. Moreover, the concentration of large numbers of poorly educated individuals in certain communities will magnify and exacerbate all other socioeconomic differences.

The divergence in education levels is causing an equally striking divergence in wage levels. Measured in 2010 dollars, the salaries of college graduates in Boston and San Jose have grown by more than $30,000 since 1980. At the other end of the spectrum, the salary of college graduates in Flint has actually declined over

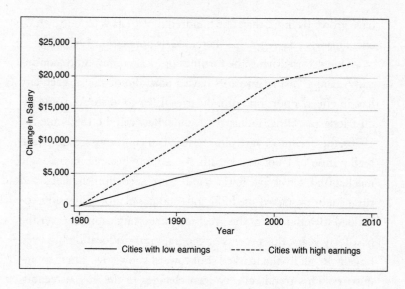

Figure 6. Gains in earnings of college graduates since 1980

this period by $11,645. Flint may be an extreme case, but the trend can be seen nationwide. Indeed, Figure 6 shows the gains since 1980 in the salary of college-educated workers for the ten cities with the highest and lowest salary levels in each year. As you can see, the gap continues to grow.

At its heart, the Great Divergence is driven by a structural shift in the American economy. Ever since the first European immigrants reached America's shores, the country's economic map has been constantly evolving. While its borders and natural landscape are largely immutable, the country's cities rise and fall as their fortunes change. This has always been true and it will remain true. Just consider this: although the total population of the United States has quadrupled since 1900, more than a quarter of U.S. counties have actually lost population in this period, a quarter have grown faster than the average, and the top twenty counties have grown by more than one hundred times. Clark County, which encompasses Las Vegas, has increased its population 1,400-fold.

The factors driving the fortunes of local communities continually change. Just as investors have a portfolio of stocks, each city has, in effect, a portfolio of industries. Like good stocks, some industries grow. Other industries decline. Between the 1880s and the 1920s, the decline of agriculture caused an enormous geographical reallocation of labor and wealth. As farming became increasingly mechanized, fewer and fewer hands were needed in the field, and rural counties started shedding jobs and population. This shift coincided with the rise of the great manufacturing capitals. Over the past forty years, this process of geographical reallocation has been largely driven by the marked shift toward knowledge-intensive industries. This trend reflects deep changes in the global technological landscape and the United States' comparative advantage in the world economy and is therefore unlikely to go away anytime soon. It is almost as if, starting in the 1980s, the American economy bifurcated. On one side, cities with little human capital and traditional economies started experiencing diminishing returns and stiff competition from abroad. On the other, cities rich in human capital and economies based on knowledge-intensive sectors started seeing increasing returns and took full advantage of globalized markets.

Much of the current debate on inequality in the United States focuses on the class divide between the America of the privileged—those with a good education and solid professional jobs—and the America of the underprivileged—those with low levels of schooling who often live from paycheck to paycheck with no job security. This view reflects the intuitive notion that technological change and globalization benefit one group and hurt the other, but it misses the important point that the two groups are affected differently in different places. Technological change and globalization result in more employment opportunities for a low-skilled worker in a high-tech hub but fewer opportunities for a similar worker in a hollowed-out manufacturing town. What divides America today is not just socioeconomic status but also geography.

The Unequal Distribution of Death

America's Great Divergence is caused by economic forces. But it is having profound effects outside the economic realm. The labor market differences between the three Americas have become so large that they are now generating a growing divide in many other aspects of our private and public life. We'll look here at four striking examples: health and longevity, family stability, political participation, and charitable giving.

Life expectancy is one of the best available measures of people's health and overall well-being. It reflects not just genetics but lifestyle, economic circumstances, and many other factors. Map 4 shows just how different male life expectancy is across U.S. counties. The East and West Coasts, together with parts of the northern Plains, tend to have higher than average life expectancy, while the South and Appalachia tend to have lower than average life expectancy. And even within each region, there is wide variation.

These differences are not surprising, per se. There is no country in the world with the same life expectancy in all of its regions. But what is striking about the United States is the magnitude of the differences. Male residents in counties with the longest life expectancies—Fairfax, Virginia; Marin and Santa Clara (where most of Silicon Valley is located), California; and Montgomery, Maryland—tend to live until they are about eighty-one years old. By contrast, male residents of counties with the lowest life expectancies tend to die at age sixty-six. In other words, the typical man in Fairfax lives *fifteen years longer* than the typical man in Baltimore,

65.9–72.1 years
72.2–73.8 years
73.9–75.0 years
75.1–76.1 years
76.2–81.1 years

Map 4. Male life expectancy, by county

just 60 miles away. The gap for women is equally large. This degree of geographical inequality in life expectancy is truly staggering, and a comparison with other countries indicates that it's substantially larger than what we see in Canada, the United Kingdom, and Japan, presumably because the economic divide between our communities is deeper than that in other countries.

Incredibly, a county like Baltimore has a life expectancy well below that of developing countries such as Paraguay and Iran. Indeed, if the bottom 10 percent of U.S. counties comprised a stand-alone country, that country would have a male life expectancy of 69.6 years and would appear very low in international rankings, squeezed between Nicaragua and the Philippines, well below China and Mexico. By contrast, if the top 10 percent of U.S. counties were a stand-alone country, it would rank near the top of international comparisons, just below Japan and Australia. (The United States as a whole ranks thirty-sixth. Although Americans spend twice as much on health care as residents of other industrialized countries, their average life expectancy is significantly lower than that of people in many other rich countries.)

Probably the most remarkable fact about life expectancy trends in America is that the vast geographical differences are not fading with time. Instead, they increase with every passing year, reflecting and possibly exacerbating the effect of growing socioeconomic differences. When David Breedlove, the Silicon Valley engineer mentioned in the introduction, moved from Menlo Park to Visalia in 1969, life expectancy in the two communities was comparable. Today life expectancy in San Mateo County, where Menlo Park is located, is almost six years longer than in Tulare County, where Visalia is located—a remarkable change.

The rising inequality in life expectancy among American communities is shown in Figure 7, which plots gains in male life expectancy since 1987 for the ten counties with the highest and low-

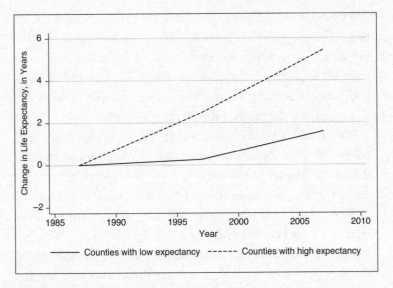

Figure 7. *Gains in male life expectancy since 1987*

est life expectancies in each year. Between 1987 and 2007, the top group gained 5.8 years while the bottom group gained merely 1.8 years. The end result is that the gap in life expectancy for the top ten counties and the bottom ten counties is much larger today than it has been in decades.*

What might be driving this stunning divergence? While access to health care for young people varies widely among U.S. counties, all individuals sixty-five and older are covered by Medicare, so differential access to care in old age is unlikely to play a major role. A more important factor is the divergence in socioeconomic conditions among different parts of the country. Education and income are among the most important predictors of longevity be-

---

* Compared with other wealthy nations, most American counties are losing ground. Since 2000 the vast majority of American counties fell in standing against the ten nations with the lowest mortality. Only 20 percent of American counties have gained ground.

cause they affect lifestyle—everything from diet and exercise to smoking and drinking habits. Thus the growing gap in education and income between the brain hubs and the rest of the country is a probable driver of the divergence in life expectancy. However, if the graph just reflected the simple sorting of highly educated individuals with high incomes into some parts of the country and less educated individuals with low incomes into other parts, it would not be particularly significant, as it wouldn't be telling us anything more than the fact that education and income drive longevity. But there is an interesting feedback phenomenon at work here that makes these findings more profound.

The geographical sorting of individuals with different educational and income levels is likely to exacerbate the longevity differences resulting from these disparities. The reason is simple: poorly educated individuals who live in a community where everyone else has low levels of education are likely to adopt less healthy lifestyles than poorly educated individuals in a community where there is a mix of educational and income levels. Economists call this a *social multiplier effect*.

For example, the probability of a person's smoking or exercising is apt to depend both on her own traits and on whether those around her smoke or exercise. The economist and former air force officer Scott Carrell has measured the importance of the social multiplier effect using data on physical fitness of members of the U.S. Air Force Academy. Members of the Air Force Academy are randomly assigned to squadrons of approximately thirty individuals with whom they are required to spend the majority of their time. The randomization of the assignment makes Carrell's data particularly useful, because it allows researchers to measure the *causal* influence of peers separate from all other possible confounding factors. Carrell and his coauthors have found definitive evidence that individuals who are assigned to a squadron where others are less fit tend to become less fit over time. The effect is very strong: poor fitness spreads like a contagious disease, with the largest effects caused by peers who are the least physically fit. The Yale economist Jason Fletcher has found similar effects for smoking. Increasing the number of smokers in a person's social network by 10 percent increases the likelihood that that person will smoke by approximately three percentage points. (As a former smoker, I can attest to the fact that the urge to light up is much stronger when I spend time in East Coast cities, where I see many people smoking outside buildings, than in California, where I rarely see people smoking.) The availability of nutritious food also varies greatly depending on the socioeconomic characteristics of each community. In low-income neighborhoods, fast-food restaurants are more prevalent and fresh food is harder to obtain than in mixed-income communities.

The social multiplier effect matters because it increases the difference in health between individuals with the same levels of income and education living in communities with different average levels of income and education. Effectively, it means that socioeconomic segregation of the type we are now seeing has an indirect effect on people's health and longevity over and above the direct effect of their own education and income. This leads to a startling conclusion: where you live has to do with how long you live.

The Moving to Opportunity program, one of the most ambitious social experiments ever attempted in the United States, is particularly interesting in this respect. From 1994 to 1998, the federal government gave thousands of public housing residents in Baltimore, Chicago, Boston, New York, and Los Angeles vouchers to leave the projects and move to private housing in the same city but in significantly better neighborhoods. Like the Carrell study, this was a randomized experiment, with 1,788 families randomly selected to receive the voucher and 1,898 families randomly assigned to the control group. When researchers visited the two groups ten years later and measured their health, their findings were astonishing. Although before the experiment the two groups were identical, the group that had received the vouchers and moved to better neighborhoods was in significantly better physical shape. Its members had improved their diet and were exercising more. They had a significantly lower incidence of obesity, diabetes, and depression. In general, they were healthier and happier, the effect being particularly strong for young women. There are many possible explanations for these findings. But a plausible one is that the place where we live and the people who surround us play an important role in shaping our health.

The Growing Divide in Divorce and Political Participation

Economic circumstances and education play an important role not just in people's health and longevity but also in their family structures. Take divorce, for example. While the factors that lead to a divorce are numerous and complex, bad economic conditions are known to be an important trigger. When things are not going well for a couple, economic troubles only make the situation worse. Not surprisingly, American cities differ enormously in their divorce rates.

What is the city with the highest divorce rate in America? If you think it's Las Vegas, think again. Using data on 8 million adults who have ever been married, I discovered that the city with the highest incidence is Flint, Michigan, where 28 percent of all adults reported being divorced in 2009. With a local economy ravaged by the closure of auto manufacturing plants, declining wages, and a disappearing middle class, Flint, together with other Rust Belt cities, has long been in a state of economic decline. Its largest employer, GM, has shrunk from a peak of 80,000 local employees to 8,000. Toledo, Ohio, another former manufacturing center, is not too far down in the divorce rankings. At the other end of the spectrum are cities like Provo, Utah, in the heart of Mormon country, where the divorce rate is low for religious reasons; State College, Pennsylvania, a university town; McAllen, Texas, with a very high density of Catholics; and Stamford, Connecticut, the best-educated and most prosperous metropolitan area in the country. San Jose is also near the bottom.

The difference in the incidence of divorce between American communities is pronounced. Flint has about three times the per capita number of divorced people as Provo. And this gap is widening. Figure 8 shows the increase in the percentage of adults who report being divorced in the ten cities with the highest and lowest percents in each year. While many cultural and religious factors play an important role, these factors are largely fixed; for instance, Provo has always been a predominantly Mormon city. The mounting divide in the figure between cities with high divorce rates and those with low divorce rates is likely to reflect the growing divide in economic conditions.

America's increased socioeconomic segregation is also affecting the political process in complex and far-reaching ways. At the national level, the balkanization of the electorate makes it harder and harder for Americans to come to a consensus on important issues that involve the future of the country. There are many reasons for this trend. Primaries are dominated by radical candidates more often than in earlier years; political coverage by cable TV networks is more and more polarized; representatives and senators face stronger incentives to vote along party lines. But geography is playing an increasingly important role. Geographical segregation raises the number of people who live surrounded by others like them, and this is likely to reinforce extreme political attitudes. In his book *The Big Sort*, Bill Bishop used data from three decades of

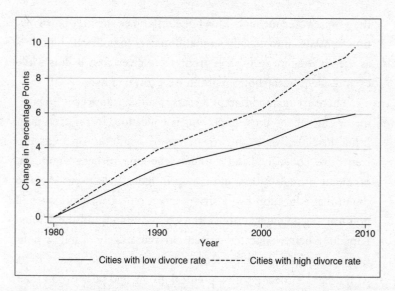

Figure 8. The change in divorce rates since 1980

presidential elections to reveal an explosion in the number of communities that are so politically homogeneous that they vote overwhelmingly for one candidate.

Surprisingly, the effect of balkanization may be just the opposite at the local level. Economists have long pointed out that it is far easier for more socially homogeneous communities to agree on local policies. For example, voters are more likely to come to a consensus on issues like local taxes, schools, parks, and police when they share similar income and educational levels—and therefore similar needs and tastes—than when they are economically very diverse.

The Great Divergence is also affecting voting patterns. American communities differ enormously in how much their citizens vote and therefore how politically influential they are. In the 2008 presidential election, the ten counties with the highest voter turnout cast four times as many votes per capita as the ten counties with the lowest voter turnout. This huge difference in voter participation translates into equally large differences in political clout. It is as if each resident in the top group were given four ballots while each resident in the bottom group were given only one.

There are many different factors that determine civic engagement. One of the most important is education. In research published in 2004, based on two surveys of 3 million American citizens, two colleagues and I found schooling to be a significant predictor of registration and voting in federal elections. We also found that education had a strong effect on broader measures of political engagement, in both the United States and the United Kingdom: better-educated citizens are more likely to follow politics in the media, be informed about the issues and discuss them with others, associate with a political group, and be active in their community. Incidentally, this is one of the main reasons both liberals and conservatives support public education. In 1962 the conservative economist Milton Friedman argued that "a stable and democratic society is impossible without a minimum degree of literacy and knowledge on the part of most citizens and without widespread acceptance of some common set of values. Education can contribute to both. Most of us would probably conclude that the gains are sufficiently important to justify some government subsidy."

Since education is such an important determinant of political participation, the increased educational polarization of the country ultimately results in increased polarization in political participation. Figure 9 shows changes in voter participation in presidential elections since 1992 for the ten counties with the highest and lowest numbers of votes per capita. The trends go up and down in accordance with nationwide turnout. The 2000 Bush versus Gore election was extremely close, and this meant that turnout

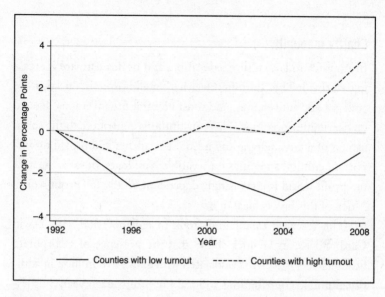

Figure 9. The change in voter turnout in presidential elections since 1992

was higher than in the 1996 and 2004 elections, when Clinton and Bush won by larger margins. The 2008 presidential election had one of the highest turnouts in recent history, probably because the nation's first African American candidate was on the ballot.

But the more interesting part of the graph is the difference between the top counties and the bottom counties. With every passing election, the top counties gain influence in the political process while the bottom counties lose it. A similar picture emerges when we tabulate financial contributions to candidates and parties, another important measure of political engagement and influence. These trends are likely to have real effect on legislation. When crafting policies that pit the interests of some communities against those of others, presidents and governors have a clear incentive to favor the needs of the communities that are politically active and better organized.

Charity Inequality

In America today, culture, education, and health care are increasingly provided by charities. Over half of all hospitals, one-third of colleges and universities, and most cultural organizations depend on charitable donations. These institutions contribute to the social capital of a city, helping the neediest residents and providing important cultural amenities for wealthier residents. In turn, the not-for-profit world is increasingly dependent on the for-profit world. Much of this link is local in nature.

The research that I undertook in collaboration with David Card and Kevin Hallock shows that the presence of a corporate headquarters in a city is associated with about $10 million in additional public contributions to local nonprofits each year. Surprisingly, it is not the corporations themselves that are driving this generous giving. Rather, it is their highly paid corporate executives. We have found that the addition of a new headquarters in a city significantly increases the number of highly paid individuals. Their salaries are often tied to their company's performance, and when their company is doing well, they tend to give generously to local charities. We estimate that for each $1,000 in market value for the firms headquartered in a city, about $1 goes to local nonprofits. By contrast, corporate donations do not appear to be very important for local charities. In retrospect, this makes sense. With customers and employees scattered across the country, large corporations have limited incentive to contribute to local causes.

The same cannot be said of individuals. When Microsoft moved to Seattle, it did not just reshape the labor market of the city, it also changed its nonprofit sector. Cofounder Paul Allen alone has donated more than $1 billion, 60 percent of which has gone to local charities to finance, among other things, the building of two new museums—the Flying Heritage Collection and the

EMP Museum—and a new library at the University of Washington, the restoration of the historic landmark movie theater Cinerama, and the expansion of the University of Washington Medical School.*

Which American cities have the most charities per capita? Among large U.S. metropolitan areas, charities in five brain hubs—Stamford, Boston, Raleigh-Durham, Washington, D.C., and New York—receive the highest contributions relative to their population. Seattle is not far behind. The difference between communities with a strong nonprofit sector and those with a weak one has been growing, both in terms of per capita number of not-for-profit organizations and per capita contributions. In other words, the Great Divergence across American communities is causing a corresponding divergence in the resources available to local charities. Cities with many corporate headquarters and a prosperous local economy are the very cities that generate the most charitable contributions for local nonprofits. By contrast, cities with few headquarters and struggling local economies—arguably the ones that need charities the most—are the ones that attract the fewest contributions. This further magnifies the distance between winners and losers.

The growing differences in life expectancy, divorce, political participation, and charitable contributions are just examples of the many ways in which American communities have been diverging over time. They are not the only ones. Many other aspects of American society have followed similar dynamics. Crime, for example, exhibits diverging trends, with cities such as New York and

---

* The other notable Microsoft charitable offspring is the Bill and Melinda Gates Foundation. Since it mainly focuses on developing countries, it plays a necessarily smaller role in supporting local causes, although it has given grants for education reform in the United States.

Boston experiencing large improvements over the past two decades and cities such as Flint and Detroit experiencing smaller improvements. At the end of the day, all these social differences are a stark reflection of differences in economic conditions. American communities have always differed from one another, some richer and some poorer. But the economic distance between the communities at the top and those at the bottom is larger today than it was fifty years ago, and it is affecting all aspects of life.

In order to begin to address these enormous shifts—and the challenges they present to our society—we must come to terms with the fundamental economic causes of the divergence. Why are some cities magnets for high-paying jobs and healthy, skilled workers while others are not? What is so special about brain hubs? And why is the gap between the hubs and the rest of the nation growing every year? As we are about to discover, the Great Divergence is not a historical accident but the inevitable consequence of far-reaching economic forces.

4

◆

FORCES OF ATTRACTION

A T FIRST GLANCE, the geographical location of America's innovation hubs appears arbitrary and puzzling. In many traditional
industries, location is tied to natural resources. The U.S. oil industry clusters in Texas, Alaska, and Louisiana because that's where
large oil reserves are. The wine industry is mostly concentrated in
California because of good weather and favorable soil. The lob-
ster industry is in Maine because lobsters don't live in Kansas. In
these cases, clustering is neither surprising nor particularly revealing. The geographical concentration of innovative industries, however, appears much harder to explain. There are no obvious natural advantages to explain why innovative industries are located
where they are. After all, there is no silicon in Silicon Valley. In
the past, firms established themselves near their customers because
transportation costs were high. For example, during the industrial
revolution, London companies could deliver their products at a
cost advantage because most of their customers were in London.

But today transportation costs are low, especially in high-tech industries. In the case of IT, they are essentially zero, since you can ship software code instantly and cheaply through any modem. If Google moved to Visalia—or to Tibet, for that matter—no user would ever notice.

Thinking a little more deeply about the location question makes it even more baffling. Companies appear to locate in absolutely the worst places: they pick very expensive areas—the Bostons, San Franciscos, and New Yorks of the world. With sky-high wages and office rents, these are among the costliest places in America to operate a business. We would expect these cities to be unattractive for firms, especially those that compete globally.

Why do innovative firms cluster near each other in these expensive locations when they could be anywhere? What is so special about cities like San Jose? If you actually visit San Jose, you will not find it very different from other cities. Visual clues as to why salaries should be so high are hard to find. In fact, there is no "there" there. Most of the iconic Silicon Valley companies are located in anonymous office buildings or office parks. Like many other metropolitan areas in the United States, the San Jose metro area is made up mostly of parking lots, corporate campuses, and a few sterile-looking glass towers surrounded by an ocean of single-family homes. There is nothing distinctive about its urban form; freeways crisscross its vast expanse, and people drive everywhere. The city has been trying to redevelop its downtown and turn it into a more pedestrian-friendly destination, but it is an uphill battle. Among the tallest buildings, the Adobe tower stands out because it is visible from the 101 freeway. Situated just 3 miles southwest is eBay's sprawling campus. Just a short drive away, you find Intel, Cisco, Yahoo, and countless less prominent high-tech corporations with arcane names that sound like prescription drugs—Progent, Xilinx, and Sanmina. To hire a skilled worker with a college education, eBay and Adobe must pay that worker $87,033 a year in San Jose, but a similar company based in Merced would need to pay only $62,411. In fact, if eBay and Adobe moved to Merced, they would end up paying less to hire a college graduate than what they are paying now to hire a high school graduate, which is $68,009. This seems truly puzzling.*

If San Francisco Does Not Like Walmart, Why Does Walmart Like San Francisco?

The story of Walmart sheds some light on the matter. The company name is synonymous with low prices, so it is perhaps not surprising that ever since Sam Walton founded the company fifty years ago, Walmart has been based in Bentonville, Arkansas. Bentonville is a small town where running a business is remarkably affordable, embodying the company's frugal corporate culture. Office space in Bentonville is among the cheapest in the nation, and the cost of living and average wages are also low. This is where Walmart's CEO, all its top managers, and all the headquarters staff live. Bentonville is clearly the perfect fit for Walmart's cost-cutting ethos.

But when Walmart set out to enter e-commerce twelve years ago, it did not choose to locate its Internet division, Walmart .com, in Bentonville. Nor did it choose Bangalore, where costs are

---

* Many industries display some degree of geographical agglomeration. Entertainment is concentrated in Los Angeles, and the main producers of carpets have long been in Dalton, Georgia. But the degree of concentration of the innovation industry is particularly high. The ten counties with the greatest number of jobs in computer manufacturing account for 70 percent of the total number of jobs. For scientific R&D, software, and Internet companies, the corresponding figures are 45, 32, and 25 percent, respectively. The majority of all nanotech research done in the world occurs in eight metropolitan areas. Countries such as Japan, France, and Korea exhibit even more geographical concentration, as Tokyo, Paris, and Seoul absorb the lion's share of their nation's most innovative activities and talent. The same is true in China, with Shanghai and Beijing.

even lower. Instead it chose Brisbane, California, just 7 miles from downtown San Francisco, one of the most expensive labor markets in the world. (It also happens to be an area that is politically hostile to Walmart, which makes it hard for the company to open many local stores.) What sense does this make, given how aggressive Walmart is in keeping the costs of every division under control? Has Walmart betrayed its own business model?*

No. As it turns out, in the world of innovation, productivity and creativity can outweigh labor and real estate costs. Walmart saw three important competitive advantages to a San Francisco location, which economists refer to collectively as the *forces of agglomeration:* thick labor markets (that is, places where there is a good choice of skilled workers trained in a specific field), the presence of specialized service providers, and, most important, knowledge spillovers. Although not much discussed, these forces ultimately determine the location of innovative workers and companies and therefore shape the future of entire communities. Understanding these forces is critical, because they are the ones responsible for the Great Divergence of the past thirty years. Understanding them is also important because they hold the key to making struggling cities more economically successful. As we'll see, they explain why eBay and Adobe find it profitable to remain in San Jose, and the seemingly illogical location decisions of many other companies, from Pfizer to IBM. These forces are growing stronger, and they will affect each and every American worker in the years to come.

---

* Actually, according to a current employee, Walmart did try to locate Walmart .com in Arkansas at first. The results were so poor—presumably because of the lack of specialized Web designers in Bentonville—that the company had to move the division almost immediately. The employee (who prefers to remain anonymous) told me that the graphics of the pilot Web page were so bad that insiders still joke about it.

Advantage 1: Size Does Matter

In 2007, Mikkel Svane cofounded the high-tech firm Zendesk in Copenhagen, but he soon realized that Copenhagen was too isolated. Two years later he moved the company to the United States "in search of [funding and] talented staff." First he tried Boston, but eventually he settled on San Francisco. "It is very exciting. Coming to San Francisco and working with the local people here and our advisers has made us think bigger and more aggressively and really pushed the envelope," Svane told the local press when he moved.

The twenty-four-year-old programmer Kiel Oleson moved from Lincoln, Nebraska, to San Francisco in 2010, looking for a high-tech job. His story is pretty typical. "I knew I wanted to do iPhone development, and a lot of start-ups here are looking for iPhone developers."

If you ask San Francisco CEOs why they moved their companies to the Bay Area—even when there are much cheaper places to do business—most will tell you that San Francisco is where the talent is. If you ask software engineers why they moved to San Francisco, they will tell you that that's where the jobs are. Simple, right? Not really. True, a software engineer will find many more job openings in San Francisco than in Lincoln. But he will also face more competition for these jobs. The same is true for employers. Although there are more software engineers looking for jobs in San Francisco than in Copenhagen, there are also more employers seeking to hire them. It is odd that workers claim that San Francisco has an excess demand for software engineers while employers claim that San Francisco has an excess supply of engineers. They cannot both be right.

The reality is that in most cities, supply and demand for specific occupations are generally well balanced. If an overabundance of software jobs creates an excess demand for software engineers in a particular city at a particular time, engineers from all over the country will flock to that city and even out demand and supply. Software engineers tend to be young, well educated, and foreign-born—three groups with particularly high mobility. During the dot-com boom of the late 1990s, there was an enormous demand for high-tech workers in Silicon Valley as hundreds of well-financed startups scrambled to hire new employees, and hundreds of thousands of high-tech workers moved there from all over the United States. During the bust that followed, demand collapsed, and hundreds of thousands moved away. But if the ratio between employers and engineers is the same in San Francisco, Lincoln, and Copenhagen, what attracted Kiel Oleson and Mikkel Svane to San Francisco? The answer is that in San Francisco the labor market for software engineers is *thicker*. This turns out to have enormous implications for cities.

In the case of labor markets, as in many other aspects of life, size does matter. Economists have long understood that thick markets—those with many sellers and many buyers—are particularly attractive because they make it easier to match demand to supply. To see why thickness is a good thing, forget about jobs for a second. Think instead about love. Imagine that you are single and looking for a partner. After several unsuccessful attempts to navigate the local bar scene, you decide to try an online dating site. Suppose there are two sites for local singles in your area, identical in every respect except size. The first site is called thick.com and typically contains the postings of one hundred men and one hundred women. The second site is called thin.com and features the postings of ten men and ten women. Which one do you choose? The ratio between men and women is the same in the two sites; in both cases, for each "seller" there is one potential "buyer." So you might be tempted to conclude that the sites are equivalent. But of course they are not. The probability that a woman will find exactly the man she is looking for—the one with the looks, interests, and values that she really wants—is higher in the larger site because there are more men to choose from. And of course the same is true for the men.

Labor markets are very similar to dating sites. Thick labor markets are better at matching employers with workers, and the ultimate match is closer to the ideal match. If you are a molecular biologist who specializes in a particular branch of recombinant DNA technology, it is really important to find the one biotech firm that uses that specific technology. If you move to a city like Boston or San Diego, where many biotech firms are concentrated, you are more likely to find the firm that really wants—and will pay for—your unique skill set. If you move to a city like Portland or Chicago, where there are fewer biotech firms, you may have to settle for a less ideal match and therefore a lower salary. You will have a vastly different career trajectory depending on where you decide to move.

The advantages of a better match also accrue to employers. By locating in Boston or San Diego, a biotech startup enjoys higher productivity and produces more patents, because it can find exactly the kind of molecular biologists that fit its needs. This eventually results in higher profits and a more successful IPO. When the startup Trulioo, which develops online identity-verification technology, moved from Vancouver to Silicon Valley, its CEO, Stephen Ufford, noticed that the company's productivity skyrocketed. "The speed at which things move [in the Valley] is quite key," he said, adding that he accomplished in three months what would have taken years back in Vancouver. A thick market is a win-win for workers and firms alike.

The economic return of being in a thick labor market, as measured by increased earnings, is significant for professionals and has been rising for the past thirty years. For example, in the United States the average wage in labor markets with over a million workers is a third higher than the average wage in markets of 250,000 workers or less. This remains true even after worker seniority, occupation, and demographics are held constant. Remarkably, this difference is now 50 percent larger than it was in the 1970s. Market size is particularly important for workers with highly specialized skills, such as high-tech engineers, scientists, mathematicians, designers, and doctors. For example, studies have shown that among doctors, specialists perform a narrower set of activities in large cities than in small ones. But market size does not matter very much for unskilled workers; manual laborers and carpenters perform similar tasks in large and small cities.

Think about the history of Facebook. As everyone who has seen the movie *The Social Network* knows, Mark Zuckerberg founded Facebook in his dorm room at Harvard University, in Cambridge, Massachusetts. Cambridge boasts the most impressive concentration of top universities in the world and is one of the best-educated cities in America. Plenty of innovative companies are in the area, and there is definitely no shortage of talent. However, Zuckerberg quickly realized that in order to find the right talents for his emerging business, he had to move his company to Silicon Valley. In the Valley, the market for engineers is so thick that Zuckerberg found not just well-educated workers, not just well-trained engineers, but well-educated, well-trained engineers with the precise skills he needed. Viewed in this light, Zuckerberg's choice of location and the general tendency for high-tech companies to establish themselves near other high-tech companies is not surprising. In isolation they would probably struggle, but by clustering close to each other, they become more creative and productive.

The size of labor markets also affects how frequently people change their jobs. One study that followed 12,000 workers over a twenty-year period revealed that early in a career, when a worker is shopping around for a good match, she will change jobs more often in a large local market than in a small one. Later in a career, however, when stability presumably becomes more appealing, people change jobs less often in large markets, because they are more satisfied with their matches.

In addition, market thickness provides a form of partial insurance against unemployment. When a layoff is caused not by a recession but by problems specific to a particular firm, market thickness reduces the probability that the worker will remain unemployed, because there are more potential employers. At the same time, thick labor markets reduce the likelihood that a firm can't fill a vacancy.

The importance of market size is not just a curiosity. It has critical implications for the future of cities. The thick-market effect is one of the main reasons for the concentration of the innovation sector in a small number of cities worldwide. Because thick labor markets are better at matching workers and employers, innovation clusters have an enormous advantage in attracting even more high-tech employers and workers. The flip side, however, is that cities that do not already have an innovation cluster find it hard to create one. Eventually this exacerbates the distance between brain hubs and all other places.

Market thickness has some interesting and unexpected consequences. For example, firms and workers joining a cluster enjoy private benefits in terms of higher productivity. But they also generate a benefit for all the other firms and workers in the cluster, which are made more productive by the new entrant. This externality is another example of market failure, a case where government intervention could improve efficiency by subsidizing workers and firms for the benefits they generate.

Surprisingly, the size of labor markets affects not just workers' productivity but also their ability to find romantic partners. For the baby-boom generation, small towns had a particular allure. But for younger Americans, the attractiveness of small towns is declining, and Generations X and Y are settling in great numbers in large cities. This trend in part reflects changing cultural norms and the increasing economic benefits of thick labor markets. But it also reflects the changing calculus of finding a mate. The marriage market in the United States has become increasingly segregated along educational lines, with well-educated professionals increasingly marrying other well-educated professionals. Economists have a decidedly unromantic term for this trend: *assortative mating*. It means that people tend to marry people with similar socioeconomic characteristics. Assortative mating is nothing new: even in the 1980s, well-educated women were more likely to marry well-educated men than less educated men. However, this tendency has strengthened over the past thirty years, with a significant increase in the probability that a man with a master's degree will marry a woman with a master's degree, a man with a college degree will marry a woman with a college degree, and so on. This applies not just to educational levels but also to type of job, salary level, and many other factors. Like attracts like. As assortative mating increases, the need for a large marriage market also increases. If you are looking for a partner with very specific characteristics, a thick dating scene is better.

Remarkably, even for married couples the need for thick cities is increasing, because large labor markets are particularly important for families in which both the husband and the wife have a professional career. This kind of "power couple" represents a small but growing number of households. In a recent study of changing family structure, the UCLA economists Dora Costa and Matthew Kahn found that in 1940, among couples in which both husband and wife had a college education, only 18 percent of wives worked. By 1970 that number had risen to 39 percent, but most of the women had majored in fields such as education and nursing and had been tracked into traditionally female jobs. Their experience was one of "first family, then jobs," and they often left the labor force when their first child was born. By 1990 college-educated wives of college-educated men had begun to resemble men in their professional choices and aspired to "career, then family" or "career and family." By 2010, 74 percent of college-educated wives were in the labor force, with jobs in virtually all fields and sectors.

As more married couples have two careers, more of them face a location problem, since wives are increasingly unwilling to be the ones who give up their careers to accommodate their spouses' job changes. Today, half of companies list a spouse's employment as the biggest reason that employees turn down a job relocation offer. Thick labor markets—markets large enough to offer good professional matches for both partners—are the best way to solve this problem.

As it turns out, this matters tremendously for the future of cities. Costa and Kahn find that well-educated professionals are increasingly gathering in large cities and that more than half of this increased agglomeration is due to the growing severity of the location problem of power couples. This is bad news for small cities, because it means that they are losing competitiveness, especially in the eyes of professional couples with high levels of education. In the long run, smaller cities are destined to experience a reduced inflow of well-educated professionals and will therefore miss out on the growth of the innovation sector, thus becoming poorer over time.

Advantage 2: Ecosystems and Venture Capitalists

Gonzalo Miranda is a Chilean venture capitalist. His firm, called Austral Capital, identifies promising high-tech startups in Latin America and brings them to the United States. I was particularly interested in meeting him because he is doing exactly the opposite of what we might expect: he is moving companies from less developed countries with low operating costs to high-cost locations in California and Washington State. The companies that he finances and helps move north end up creating good jobs in America. Since he focuses on early-stage startups, the companies are small when they arrive, but he estimates that the typical company creates about twenty jobs in its first year in the United States, mostly in engineering and support staff. If it succeeds, that number can easily grow to hundreds or thousands.

Miranda tells me that the entrepreneurs from Chile, Argentina, and Brazil that he works with are typically brilliant innovators with world-class new technologies. Their products are so good that even if they stayed in their home countries, they would probably do well. But if they want to make it big, they must move to the United States. One advantage of America's innovation hubs, he says, has to do with funding, which is scarce in Latin America and dominated by old-style investors who often end up suffocating a company. Another advantage is the legal system. Even in Brazil and Chile, which are among the most pro-business countries in South America, setting up a business can be difficult, and the entry costs are uncertain, owing to burdensome bureaucracies and complicated legal requirements. But above all, moving to Silicon Valley or Seattle means having access to an entire ecosystem. "The Silicon Valley ecosystem ends up being an advantage that more than compensates for the higher costs," Miranda told me.

The ecosystem that Miranda is referring to includes providers of specialized services that are important to innovative firms, such as advertising, legal support, technical and management consulting, shipping and repair, and engineering support. These services enable high-tech firms to focus on what they are good at—innovation—without having to worry about secondary functions. By the mere act of moving into a high-tech cluster, a firm in effect becomes larger overnight, because it can draw on specialized local expertise. As a result, high-tech firms within a cluster become more productive and more successful. A small software developer in Seattle does not need an in-house lawyer, because there are already plenty of local law firms specializing in intellectual property, licensing, and incorporation of startups. A biotech company in Durham can buy specialized services for its labs from local vendors, and a hardware company can find specialized shipping services. A young entrepreneur in Silicon Valley told me that he needed to have his startup incorporated but did not want to pay thousands of dollars in legal fees. In any other location, he might have given up on starting his business, but because he is in Silicon Valley, he easily found a law firm that accepts equity instead of cash. This firm's business model, which rests on the assumption that one of the hundreds of startups it incorporates will turn into the next Google, is possible only in a dense high-tech cluster.

From the point of view of the providers of these specialized services, geographical proximity to clients is crucial. They need to be close to potential clients to assess their needs and show how they can help. This matters less for mature products, but it is critical when a product is completely new. Cadence Design Systems is a leading high-tech company that makes the software other firms use to design electronic systems. Its clients include IBM, Siemens, NVIDIA, Silicon Laboratories, and Casio, most of which have a presence in Silicon Valley. When I asked Chi-Ping Hsu, the company's senior vice president for R&D, why Cadence keeps most of its R&D workforce in San Jose, he replied, "How do you determine what the architecture of the new product should look like? You go to the customer every other day with a prototype."

This is one important factor that keeps the ecosystem geographically together: Cadence is in Silicon Valley because its clients—other high-tech companies—are there. And the clients are there because their vendors, including Cadence, are there. If this sounds familiar, it is because we saw exactly the same thick-market effect with workers and employers. This affects local communities in two important ways. First, it increases the number of local jobs created by high-tech firms. If a city attracts an IBM office, it gains not just the IBM jobs but also the jobs at Cadence, as well as jobs at all the other service providers. This is one reason that the high-tech multiplier is large. Second, it further strengthens the attractive power of cities that have a significant innovation presence at the expense of cities that don't: it makes IBM more likely to open an office in Silicon Valley.

Take Ericsson, the Swedish phone giant, which is technically headquartered in Stockholm. The *Wall Street Journal* reports that Håkan Eriksson, the company's chief technology officer, doesn't have an office in Stockholm. Instead he works from San Jose, "where Ericsson has more than 1,200 employees engaged in R&D." In the past Ericsson benefited from its proximity to Nokia, the Finnish cell phone giant. But now the company finds it more important to be near Apple, because of the iPhone and iPad, and near Google, because of the Android operating system for cell phones. "The epicenter for the handset industry has shifted from Finland to Silicon Valley," Mr. Eriksson said. Incredibly, even Nokia seems to agree; it now has a research center in Palo Alto that employs 380 workers, 80 of whom hold PhDs. Director John Shen told the *Wall Street Journal*, "To be globally competitive, you really need to have a footprint here in the Valley."

Possibly the most important part of the high-tech ecosystem is venture capital. After the subprime fiasco and the Great Recession, financial innovation developed a bad reputation. But a strong financial system is crucial for job creation. Contrary to general perception, a strong financial system is less important for the rich, who already have money, than for those who are trying to get it. Venture capital is a brilliant solution to an old problem. Young people tend to have creative new ideas but often lack the capital to realize them. The job of a venture capitalist is to identify, among the thousands of new ideas, the ones that have promise. It is a very democratic concept and a crucial piece of the American Dream.

I grew up in Italy, a country where financing for innovative ventures is hard to come by. During the Renaissance, Italian banks, especially those in Florence, were market leaders in Europe, but in more recent centuries they have fallen behind. As a consequence, it is typically quite difficult for a young Italian with a bright new idea to start a business in high tech. This constraint creates two very costly distortions. First, it tends to dramatically reduce innovation: Italy has an underdeveloped high-tech sector despite its top-notch engineering schools. Second, the lack of venture capital financing is terribly unfair, because it discriminates against those who come from disadvantaged backgrounds. If you have a good entrepreneurial idea and come from a well-to-do family, you might get financing, either directly from your family or because your family can provide collateral to the bank. But if you have the same idea and come from a poor family with no collateral, you're probably out of luck. It is a tremendous waste.

In 2010, I decided to take a sabbatical from my job at the University of California at Berkeley to spend the year at Stanford as a visiting professor. Every morning on my way to work, I would drive along Sand Hill Road in Menlo Park. Sand Hill Road contains the largest concentration of venture capital firms in the world.

All major VC firms are located there, including the mythical Sequoia Capital and Kleiner Perkins Caufield & Byers, early backers of the most iconic startups in the history of high tech: Google, Apple, Amazon, Oracle, Yahoo, YouTube, PayPal, Netscape, and Cisco. I often saw young entrepreneurs with big dreams entering one of those low-rise buildings, presumably to pitch their ideas. Those low-rises on Sand Hill Road are where venture capitalists determine the future of commercial innovation. The aspect of the VC industry that I find most remarkable is how local it still is. Venture capitalists used to talk about the "twenty-minute rule": a venture capitalist only considers financing companies that are located within a twenty-minute drive of his office. Today the industry has become more global, but it still has a predilection for local ventures. One study finds that the likelihood of financing new ventures declines quickly with the distance between the VC company and its target.

This comes as no surprise to Shelby Clark, the founder of RelayRides, an innovative car-sharing startup that enables car owners to rent their vehicles to other users. Clark reportedly had to move his startup from Boston to San Francisco in 2011 "to be closer to the company's backers, August Capital and Google Ventures, a division of Google Inc." Jose Luis Agell, who leads the business development of a Spanish startup called 3scale Networks, based in both Barcelona and California, says that "it is tough to get funding as a foreign company." People have remarked for decades that one of the secrets of the success of Silicon Valley is its deep and articulated venture capital base. But what does proximity have to do with funding? Why, in a world of fast communication and cheap plane tickets, should venture capitalists on Sand Hill Road favor startups near them?

Bill Draper has the answer. He is one of the most experienced venture capitalists in Silicon Valley, with forty years on the job. In his view, money is only one of many things that venture capitalists provide to startups. "There's a lot of support, a lot of team building, a lot of organization, and relationships between entrepreneurs and venture capitalists that are key in making a successful startup," he said in a recent interview. Today venture capitalists do not simply write a check and then disappear. An increasingly important part of their job involves active monitoring, nurturing, and mentoring of new businesses. This is why location matters.

Nurturing and monitoring are clearly easier if the startup in question is nearby. It would be much harder for a venture capitalist in Silicon Valley to monitor the progress of the brilliant but poor Italian entrepreneur, however exciting his idea may be. Google may be the most famous example of the importance of mentoring in the history of venture capital. Among early investors in Google was John Doerr, from Kleiner Perkins Caufield & Byers. His early financing was critical to helping Google survive its early years. But even more important was his insistence that Sergey Brin and Larry Page—brilliant engineers, but at the time still naive businessmen—hire an experienced executive as CEO. Doerr's guidance led the two founders to pick Eric Schmidt, arguably one of the most defining business decisions Google made in its early years. The Kleiner Perkins Caufield & Byers office is only 10 miles from Google. The search for a suitable CEO took almost a year and there were many dead ends, mostly because Brin and Page were reluctant to follow Doerr's advice and had to be continually prodded by him. It was like searching for a spouse—very difficult from a distance.

San Francisco–based i/o Ventures finances early-stage high-tech startups and is part of a wave of VC firms that spend a significant amount of time and energy on mentoring. Local high-tech legends such as Russel Simmons of Yelp provide counseling and business advice. Crucially, one of the requirements the company imposes on the firms it finances is that they move to San Francisco. It argues that the benefits from this kind of startup accelerator program—in which the venture capitalists guide firms from idea to product launch—can occur only face-to-face.

In the end, geographical proximity to venture capitalists still matters. Skype and cell phones have not changed this simple fact. This is one of the reasons that the world of high tech is and will remain geographically concentrated.

Advantage 3: The (Almost) Magical Economics of Knowledge Spillover

ECOtality is a company at the forefront of clean transportation and power storage technology, which in 2010 moved its headquarters from Arizona to the Bay Area. It is not the only one. The German company Q-Cells, the Chinese companies Trina Solar, Suntech, and Yingli Green Energy, and the Spanish company FRV have all recently opened shop in the Bay Area. Clean-tech companies are increasingly locating their headquarters and/or their R&D labs in the region. In a recent interview with the CEO of ECOtality, NPR reported that the reason for his move was "to be close to the action." On some level, this makes intuitive sense. After all, who wants to be away from the action? But on a deeper level, it raises the question of what "being close to the action" really means. Why would all these companies want to be so close to their competitors? What advantages could they possibly derive?

The answer has to do with a simple fact: new ideas are rarely born in a vacuum. Research shows that social interactions among creative workers tend to generate learning opportunities that enhance innovation and productivity. This flow and diffusion of knowledge represents a crucial third advantage for workers and firms that locate within an innovation cluster.

As we saw earlier, average education is the most important reason that wage levels vary so much across cities. The fact that workers earn more in brain hubs is not an accident but a reflection of the higher labor productivity that comes from working alongside highly skilled colleagues. The wage of the same person can be vastly different depending on whether the people who live around him are well educated or poorly educated. This makes sense in some general way, but it does not tell us why it is true. Understanding exactly how and why knowledge diffuses within a circle of friends, colleagues, or scientists in a city is vital if we want to understand why innovative industries locate in some cities but not in others. Paul Krugman, who conducted trailblazing academic research in this area before becoming a *New York Times* commentator, once famously wrote: "Knowledge flows are invisible; they leave no paper trail by which they may be measured and tracked, and there is nothing to prevent the theorist from assuming anything about them that she likes."

Krugman's skepticism motivated many researchers to intensify their efforts to measure the diffusion of ideas. In 1993 three economists—Adam Jaffe, Manuel Trajtenberg, and Rebecca Henderson—found a usable paper trail: patent citations. When filing for a patent, an inventor is required to list all the previous inventions that her invention builds upon. These links offered the economists an innovative way to track the flow of knowledge among inventors. Their results were surprising: knowledge is subject to a significant degree of "home bias," in the sense that inventors are significantly more likely to cite other inventors living nearby than inventors living farther away. Because patents are freely available, citations shouldn't necessarily display geographical favoritism. An inventor in, say, Durham, North Carolina, should have the same awareness of products or ideas generated elsewhere as he does of those generated in Durham. And yet an inventor in Durham is much more likely to cite a previous invention patented by someone else in Durham than one patented in another city.

The magnitude of the home bias is substantial. Excluding citations that come from the same company, citations are twice as likely to come from the city of their citing patent than from other places. This means that scientists and inventors are more familiar with knowledge produced by those who work near them, presumably because they share ideas and information through informal conversations and interactions. These interactions take place both inside and outside the workplace, including casual settings like local cafés and social events. In Silicon Valley, for example, weekend cricket games are known to be opportunities not just for physical activity but also for networking and exchange of business-related information for the local community of Indian engineers. Increasingly, one of these engineers told the *New York Times*, "cricket spills over into business." Geography matters for the spread of knowledge, and knowledge quickly dies with distance. Citations are highest when the citing inventor is located between zero and 25 miles from the cited inventor. Citations are significantly lower when the citing inventor is more than 25 miles from the cited inventor, and the effect completely disappears beyond 100 miles.

Geographical distance seems to impede the flow of ideas even *within* the boundaries of a firm. This alone should discourage companies from outsourcing any part of the innovation phase to low-cost countries. Take the high-tech company Cadence, with about two thousand employees in San Jose, one thousand workers in India, and another thousand scattered around the world. An Indian software engineer at level T4 makes about a third of what a similarly qualified software engineer in San Jose makes. When I asked Cadence's executive senior vice president, Nimish Modi, why the company does not move more R&D to India, given the potential savings, he told me that proximity and personal interaction matter to the creativity of its engineers. "We have sophisticated videoconferencing facilities, and we use them all the time to communicate with India. But it is not the same as face-to-face interaction. Nothing replaces a group of engineers sitting together and arguing in front of the whiteboard," he said.

As an academic, I am not too surprised by this. Although I communicate daily with distant colleagues by phone and e-mail, my best ideas often occur when I least expect them—over lunch with colleagues or at the water cooler. The reason is simple. Phone and e-mail are great ways to transmit information and keep a research project going once the key creative ideas are in place, but they are not the best way to come up with those ideas. New ideas arise in mysterious and unpredictable ways from free and unstructured interactions. It would be ridiculous to schedule a phone call with a distant colleague to come up with a new idea. My guess is that most researchers share this view. After all, the reason we spend so much time in academia discussing whom to hire and fire is that our colleagues affect our own productivity.

Being around smart people tends to make us smarter, more creative, and ultimately more productive. And the smarter the people, the stronger the effect. Pierre Azoulay, Joshua Graff Zivin, and Jialan Wang quantified this by focusing on what happens to medical researchers when they work with an academic superstar. It is difficult to establish the causal relationship here because of self-selection: superstars tend to work with strong researchers, so the fact that their collaborators are especially prolific may just happen because they are better, not because they are benefiting from knowledge spillovers. To control for this, the three economists had a smart idea. They focused on what happens to the productivity of a superstar's collaborators when the superstar dies unexpectedly (they identified 112 such deaths). Although nothing changed in the collaborators' own circumstances following the superstars' deaths, they experienced "a lasting 5 to 8 percent decline in their quality-adjusted publication rates."

It is not just that people publish more when they are close; the quality of their research is better. When a team of Harvard Medical School doctors analyzed all medical research articles published at Harvard and correlated their data with the distance between the authors' offices, they found that being less than one kilometer away raised the quality of research, as defined by how many other researchers cited the article. The effect was even larger if the authors were in the same building or used the same elevator.

Thus innovative firms have an incentive to locate near other innovative firms. In the same way that having a good colleague next door affects my creativity, having good neighbors—even competitors—improves the creativity of companies and workers. This in turn helps explain why workers in brain hubs earn higher salaries than identical workers in other areas. There is something almost magical in the process of generating new ideas. By clustering near each other, innovators foster each other's creative spirit and become more successful. These effects have gained importance over time. While many people think that e-mail, cell phones, and the Internet have made physical proximity less important to the creative process, in reality the opposite is true. Location is more important than ever, in part because knowledge spillovers are more important than ever. This is a key reason for the accelerating divergence in the fortunes of the three Americas.

The growing importance of knowledge spillovers does not just affect the cities where businesses and workers are clustering. It is also reshaping the physical layout of the workplace. Offices used to be simple rooms with doors, until the open-space craze revolutionized the design of many white-collar workplaces and introduced the idea of Dilbert-like cubicles. One of the most intriguing new trends is the idea of "cowork." Born in California, the concept is spreading quickly through the entire country. Cowork spaces typically host dozens (in some cases hundreds) of entrepreneurs, innovators, and artists, who rent desks or offices in the same building and sit next to each other. They are part of a growing number of creative professionals in American innovation hubs who are self-employed and prefer to remain independent of larger companies. In cowork spaces, each of them works on his project, but the setup is appealing because it offers the possibility of sharing ideas, building connections, and fostering creativity. It turns isolated innovators into a real community, a creative ecosystem designed to maximize knowledge spillovers.

One example is the San Francisco Chronicle Building. It includes, among many other ventures, a high-tech incubator, a school of digital filmmaking, an art gallery, a tool workshop for "inventors, makers, hackers, tinkerers," and hundreds of engineers, scientists, artists, and social entrepreneurs who have decided that they have a lot to learn from each other. When you enter one of these places, you feel more like you are at a graduate school than in a regular office: people network, exchange tips on how to solve technical problems, and comment on each other's business plans. The goal is "radical collaboration" among individuals who would otherwise be isolated, working alone in their home offices or garages. In a random day at the Chronicle Building, you might find a fashion designer sketching her future hat collection next to a mechanical engineer operating a laser cutter next to a Berkeley MBA writing a grant for a new Darfur nonprofit organization. The creative energy is palpable. Some cowork places organize lectures or demonstrations of the latest technologies, such as "Startup Demo Night." Others have happy hours and brown bag lunches with angel investors. It is all about establishing connections and sharing insights. The phenomenon is so new that there are no rigorous studies of how coworking affects creativity and business success.

But all the signs point in the right direction. For example, one of the Chronicle Building's early success stories is Square, a mobile phone credit-card payment processing company started by Twitter cofounder Jack Dorsey, which has grown from five to one hundred employees in just a year.

Why the Brain Drain Is a Good Thing

There is something phenomenal about the three forces of attraction. They are responsible for turning a collection of individual workers and firms into an integrated creative commons that is much larger than the sum of its parts.

This generates what economists call *localized economies of scale*. The term *economies of scale* usually refers to the ability of companies to become more efficient as they grow in size. For example, large car manufacturers are more efficient than small ones. But instead of applying to a single company, these economies of scale apply to all the companies in a geographical area. Larger clusters are more efficient because they have a thicker labor market, a more specialized supply of business services, and more opportunities for knowledge spillover. The effect can be amazing: while individual companies in a cluster do not necessarily become more efficient as they grow in size, all companies taken together become more efficient as the cluster grows. A surprising implication is that as a country, the United States is more productive—and therefore richer—because its innovation sector is concentrated in a limited number of innovation hubs rather than spread out among all cities. This is one of the paradoxes of our knowledge economy. The forces of attraction and the agglomeration of economic activity create differences and inequality among communities. But at the same time, a significant part of America's economic vitality and prosperity depends on them.

The three forces of attraction are further magnified by the tendency of engineers, scientists, and innovators to leave established companies to open their own shops. This process of procreation exists in all industries, but it is considerably stronger in the innovation sector, in part because employees of innovative firms tend to be a very special group of people. They may be nerds to an outsider's eyes, but they are exceedingly creative and entrepreneurial. Often the very success of their employer leads them to leave. In their early years, startups tend to have an irreverent culture and a nonhierarchical work environment. But with success and growth, they inevitably become more formal and less exciting, prompting some of their most entrepreneurial employees to start their own ventures. This spawning process is often facilitated by stock options packages, which, when vested, can turn into seed money for a new business.

This tendency of a company's smartest employees to start their own ventures is often referred to as *brain drain*. Companies are aware of the risk of losing their best talent and often fight back. Intel grants sabbatical leaves. Google allows all employees to work 20 percent of the time on their own projects. When key employees threaten to start their own ventures, Google has been known to offer them the opportunity to start their own enterprises within Google. These perks, which are rare outside the high-tech world, are a testament to how crucial a worker's creativity is in the high-tech sector and how important it is for companies to retain good employees.

But what is costly from the point of view of an individual company is highly beneficial for the community as a whole, because it means more local jobs. Because of the magnetic attraction of clusters, offspring don't stray too far from their parents. Research shows that the reproductive process is rarely a zero-sum game in which the young companies gain at the expense of their elders. Instead the process ends up resulting in a net gain in employment for the local community. And it gets even better: the children in turn produce their own children. Thus, from the perspective of local governments, attracting a high-tech job today will result in many more jobs in the future.

What This Means for the Three Americas

Several implications fall out of this new understanding of the forces of attraction. The most important is that the economic performances of cities will keep diverging over time. We have seen that America's economic map is highly uneven. At one end of the spectrum are the brain hubs, with highly skilled and highly productive workers earning high wages. At the other end are cities whose workers have limited skills, low productivity, and falling wages. The divide between the different Americas is growing with every passing year, and now we know why.

This divergence is the predictable result of the three forces of agglomeration. These forces inevitably magnify the differences between winners and losers among American communities. Cities with the right sectors and with workers who have the right skills are strengthening their position, while others, trapped by their past, are losing ground. It is a tipping-point dynamic: once a city attracts some innovative workers and companies, its economy changes in ways that make it even more attractive to other innovative workers and companies. This tends to generate a self-sustaining equilibrium, with many skilled individuals looking for innovative jobs and innovative companies looking for skilled workers. In the end, this is why workers in Boston are paid twice as much as workers in Flint. As the forces of agglomeration reshape the economic map, this geographical divergence is bound to strengthen over time.

A second implication is that once a cluster is established, it is hard to move it. This is a case where the future of a city depends on its past. Social scientists call it *path dependency*. Take, for example, the aerospace industry, which has historically been located near Los Angeles. In 1993 the urban planner Ann Markusen interviewed executives at major aerospace companies, asking them why they were located in Southern California. Their answers were very instructive. As an official at Northrop put it, "If one was building the first plant, one wouldn't put it in L.A. Los Angeles would not even be in the top ten. But there is a terrible cost to moving." At TRW, the answer was even more explicit: "The cost of living here is high, the traffic situation is terrible. But we'll probably stay here." Because of the forces of attraction, it was hard for the industry to relocate, even if the conditions that made Los Angeles attractive in the first place were long gone.

This means that regions without an innovation cluster will find it difficult to start one. It is a chicken-and-egg problem. Specialized high-tech workers will not move to a city that does not have a cluster because it will be hard to find an employer that values their unique skills. High-tech companies will not move there because finding specialized labor will be difficult. This presents a terrible challenge for communities that have fallen on hard times and are struggling to reinvent themselves.

For the United States as a whole, however, the implication is more favorable: it means that America's innovation sector is to some extent protected from foreign competition. Because of the three forces, it is harder to delocalize innovative activity than to delocalize physical manufacturing. A toy factory or a textile factory is a stand-alone entity that can be put pretty much anywhere in the world where transportation is easy and labor is abundant. In comparison, a biotech lab or an innovative high-tech company is harder to export, because you would have to move not just one company but an entire ecosystem. If we were starting over, I am not sure that America today would be the most obvious candidate for the world's innovation hubs. But we are not starting over. America's innovation clusters give it an undeniable advantage over Europe, China, and India.

None of this should be an argument for complacency. The forces of agglomeration are no guarantee that we will keep our leadership in innovation forever. The lock-in effect in aerospace lasted for a while, but twenty years after Markusen's study was published, the lock-in effect is much weaker, and the number of aerospace jobs in Los Angeles is much smaller. As we are about to discover, all the United States has is a head start.

Why the Secret of Success Is Adaptation

A market economy is never static. Products that are cutting-edge today will soon become commodified and easy to make. Industries that are on the technological frontier will become mainstream and, later, relics of the past. What is a good job today will inevitably become a bad job in the future. This dynamic was first recognized by Karl Marx, who thought that it was evidence of the inherent instability of the capitalist system. Eighty years later, however, the Austrian economist Joseph Schumpeter pointed out that instead of being a flaw, this process of "creative destruction" is capitalism's greatest strength and its engine of growth.

By its very nature, the innovation sector is the part of a market economy where creative destruction matters the most. The Princeton economist Alan Blinder recently noted that in the 1950s, companies making television sets were at the heart of America's high-tech sector and generating tens of thousands of high-paying jobs. After a while TV sets became just another easy-to-make commodity, and today no TV set is made in America. The computer manufacturing industry picked up where the TV industry left off, and for a while it was responsible for 400,000 high-paying jobs. We saw earlier that most of these jobs have now moved elsewhere. But this is not a sign of failure. Indeed, it is a sign of success. To remain prosperous, a society needs to keep climbing the innovation ladder. As Schumpeter argued, it is the dynamic that has been ensuring our prosperity since the beginning of the industrial revolution.

The crucial question for America's future is therefore whether our innovation clusters can adapt and reinvent themselves to maintain their edge. Clusters, unlike diamonds, are not forever. At some point the industry that supports them matures, stops bringing prosperity, and turns into a liability. The forces of attraction provide an important advantage, but once-mighty clusters have collapsed in spectacular ways. In its heyday, the Detroit auto industry was one of the country's most important innovation hubs, arguably the Silicon Valley of its time. Like Silicon Valley today, Detroit was full of technologically superior companies that were the envy of the world. The economist Steven Klepper has shown that to an astonishing degree, the rise of Silicon Valley has been tracking the earlier rise of Detroit in terms of population, employment, startup creation, and innovation. Thus, Detroit's remarkable trajectory holds important lessons for the future of our current innovation hubs.

Just like Silicon Valley today, Detroit used to think of its primacy as unassailable. In the 1940s and 1950s, its dominance of the auto industry appeared so strong that everyone started wanting a piece of it. Unions became increasingly aggressive in their demands for higher wages, more generous benefits, and rigid work rules. Management became complacent and began neglecting efficiency. Politicians thought that the auto industry could not move anywhere else and ignored the growing threat posed by southern states with right-to-work laws. More fundamentally, however, the city's fatal flaw—the one responsible for its ultimate demise—was its inability to adapt. Clusters can't afford to cling to a declining industry but need to leverage their unique strengths to reinvent themselves *before* the tipping point is reached and the local ecosystem enters a downward spiral. If they fail, the downfall can be swift and painful. The same attractive forces that fuel the rapid rise of clusters when things are good cause an accelerated collapse when things turn bad. Detroit's mistake was not the failure to stop the demise of jobs in auto manufacturing. Different industrial relations, management practices, and political decisions could have postponed the decline, but it was just a matter of time until auto manufacturing stopped being an engine of growth. Rather, Detroit's mistake was its failure to redirect its ecosystem into something new when it still had an ecosystem.

This may well be the defining difference between Detroit and the San Francisco–Silicon Valley region. The area continues to experiment and adapt to the ever-changing technological landscape. San Francisco was once an industrial powerhouse, anchored by a major port. In the 1970s its ecosystem steered decisively toward professional services and finance and later to high tech. This process of reincarnation keeps taking place today. In 1990 the majority of high-tech jobs in the region were in hardware. Now more than 70 percent of its jobs are in newer technologies, including Internet, social media, cloud computing, clean tech, and digital entertainment. Life science research has significantly expanded its footprint.

The secret of success in a changing world is constant adaptation. As the definition of what is high technology evolves, so does the Bay Area. Rather than clinging to one product or one way of doing things, the region reshapes itself every year. The forces of attraction anchor skilled labor and specialized services, but the exact kind of skills and services evolve over time, following the changing terrain of the technological frontier. This ensures that when good jobs turn into bad jobs, there is a wave of new jobs to replace them. In a sense, this creative destruction is the true hallmark of a successful cluster, one that leverages the forces of attraction in a dynamic way.

"All politics is local," as the former Speaker of the House Tip O'Neill said. For all its glamour, the world of innovation is even more local than politics. Different communities differ in their values and expertise, and this inevitably shapes the new ideas they generate, ultimately resulting in something unique and hard to reproduce elsewhere. The innovative process is largely about the unexpected cross-fertilization that results when different parts of a community connect. In this respect, Silicon Valley's move toward diversification is extremely important, because it deepens the complementarities and fosters the constant exchange of ideas and talent between different parts of the high-tech ecosystem. For example, the region's unique strength in both medical research and gaming has caused these two seemingly unrelated sectors to become intertwined in the form of "serious games," products that apply cutting-edge gaming technologies to cure diseases. A local company called Posit Science already produces gamelike software intended to improve memory and attention and possibly even help treat such disorders as autism and schizophrenia. Another example is Stellar Solutions, an aerospace company specializing in infrared missile warning systems and capsules to carry astronauts to Mars. Lately the company has been leveraging the expertise that makes it a leader in technologies of the heavens to develop technologies of the earth: its engineers are using electromagnetic waves to forecast earthquakes. The company's CEO predicts that one day you will "turn on your TV and see not just hurricane warnings, but earthquake warnings."

Not all of America's innovation hubs are equally successful at adapting. The landscape is dotted with cities that were once powerful engines of innovation and have failed to reinvent themselves. In the 1980s, for example, Rochester, New York, was an important cluster for innovation in optical technologies and imaging. Xerox was founded there in 1906 and retains a presence in the city to this day. Kodak is based there too, and until the 1980s it employed 62,000 workers. Back then Kodak was the Google or the Apple of its period and Rochester was one of the top producers of patents among U.S. cities. (Recall that until the mid-1990s, makers of photographic equipment and films dominated the list of patent producers.) Local wages vastly exceeded those in the state and in the nation. But with the advent of digital photography, people stopped buying Kodak films. The company never fully adjusted to the new digital landscape, and today it employs only 7,000 workers.

This was a serious blow, but it did not have to be fatal. Companies come and go, but communities don't. The fundamental problem with Rochester was that the local high-tech cluster was not able to move to something new. As in the case of Detroit, local entrepreneurs never branched out into different parts of high tech in significant numbers and the community as a whole failed to make the transition to a new productive landscape. The University of Rochester remains a major research engine, there are some new high-tech companies in town, and patent creation is still happening, but it is clear that the city's most dynamic years are in the past. Wages have fallen significantly below the state average and population is declining. Vacancies are so common in some neighborhoods that municipal workers paint murals with images of the city's glory days along vast stretches of empty storefronts to hide the eyesores, according to the *Wall Street Journal*. The area surrounding Kodak's headquarters, once bustling with activity and commerce, today looks like a ghost town.

Looking forward, the forces of attraction raise two key questions for the United States. The first is what we should do to retain, foster, and strengthen our innovation hubs. How can we, as a nation, maximize the chances that our innovation hubs follow the path of the San Francisco–Silicon Valley cluster and not that of Detroit and Rochester? The second question is how to help the many remaining cities that do not have a concentration of good jobs and skilled workers and are lagging behind.

Before turning to these questions, however, we need to understand more fully how vast economic differences between cities can persist over time. As we have seen, the differences in wages and salaries among American communities are enormous, and now we know why: they reflect equally large differences in the productivity of workers, meaning that it makes sense for innovative companies to locate in innovation hubs, even if the costs of doing business are much higher. But what about the workers themselves? If the difference in salary and living conditions is so large, what prevents them from moving en masse from weak labor markets to stronger ones? Shouldn't we all live in Seattle or Austin? We now turn to the issue of mobility and what it implies for communities across the country.

5

◆

THE INEQUALITY OF MOBILITY
AND COST OF LIVING

A<small>MERICANS HAVE HISTORICALLY</small> been an unusually mobile
people, constantly seeking better economic conditions. Yet this
process of geographical readjustment is not perfect. In fact, it is
highly uneven. Even if everyone is completely free to move to look
for a better life elsewhere, not everyone takes advantage of the opportunity. As it turns out, this has profound implications for inequality in America.

In Italy, where I grew up, most people spend their entire lives
in the city where they were born, which is often the city where their
parents were born. Young Italians are particularly immobile. In a
study published in 2006, I calculated that Italians tend to live with
their parents until quite late in life: 82 percent of Italian men between the ages of eighteen and thirty still live at home. And when
they do leave the parental nest, they do not move far away. Young
people commonly get an apartment in the same neighborhood as
their parents, often in the same building. While Italians may be an

extreme case, Europeans are generally much more geographically rooted than Americans. Compared with people in most other developed nations, Americans are outliers. The Great Recession has temporarily slowed Americans' mobility, but once the economy rebounds, people will start moving again.

This willingness to relocate is a large factor in the country's prosperity, and it always has been. Tocqueville remarked in the nineteenth century that "millions of men are marching at once toward the same horizon; their language, their religion, their manners differ; their object is the same. Fortune has been promised to them somewhere in the west, and to the west they go to find it." In the late nineteenth and early twentieth centuries, migration from rural communities to urban areas provided the crucial labor that fueled the expansion of America's mills and factories. The economic historian Joe Ferrie, one of the foremost experts on this issue, noted that migration "facilitated the exploitation of natural resources at locations distant from the narrow band of initial settlement on the Atlantic coast. Farmers moved to more fertile land in the Ohio River Valley in the late eighteenth century and on to the Great Plains by the middle of the nineteenth century. And mineral and timber resources were worked by migrants to the West and the Northwest. By the Civil War, much of the gap in wages between the West and the East in the Northern states had been erased." More than that of any other developed country, America's population has always been on the move, chasing the next opportunity. Using a detailed data set painstakingly assembled from original entries in historical censuses, Ferrie estimated that even in the nineteenth century, Americans' propensity to move was twice that of residents of Great Britain or Japan in the same period.

Today about half of American households change addresses every five years, a number that would be unthinkable in Europe, and a significant number relocate to a different city. About 33 percent of Americans reside in a state other than the one in which they were born, up from 20 percent in 1900. This staggering degree of mobility has both positive and negative effects. On the one hand, moving has social and personal costs. Americans tend to live farther from their parents and siblings than Europeans. When they have children, they are less able to rely on their family for help in raising them. They are less attached to their neighborhoods and less familiar with their neighbors. But there are also advantages to mobility: if the economic conditions in a region are not particularly good, Americans are apt to look for better opportunities elsewhere. By contrast, Italians and other Europeans tend to stay put. At the individual level, Italians are giving up career opportunities and higher salaries to be close to their parents and friends. At the national level, this immobility worsens the unemployment problem and lowers overall job and income growth. In some regions of Italy (typically in the north), there is an abundance of high-paying jobs and virtually no unemployment. In other regions (typically in the south), there are very few jobs, low salaries, and high unemployment. By not moving north, young Sicilians and Neapolitans effectively increase the unemployment in their region, a situation that leads to less prosperity and stunts Italy's potential for growth.

Although Americans as a whole have always been much more mobile than Europeans, there are large differences among them, with some groups much more willing to move than others. At the time of the Great Migration in the 1920s, when more than 2 million African Americans abandoned the South for industrial centers in other regions, less educated individuals were more likely than others to migrate in search of better lives. Today the opposite is true: the more education a person has, the more mobile she is. College graduates have the highest mobility, workers with a community college education are less mobile, high school graduates are even less, and high school dropouts come at the bottom of the list.

In this respect, American high school dropouts are more similar to Italians than to American college graduates. And it's not because of a lack of opportunities. The United States is a large and diverse nation, and it is always possible to find cities and states that are doing better than others. These geographical differences can be very large. In 2009, at the peak of the Great Recession, unemployment in Detroit reached 20 percent, while unemployment in Iowa City, about 500 miles west of Detroit, was only 4 percent. The experience of unemployed workers in the two cities could not have been more different. A 4 percent unemployment rate is so low that economists consider it effectively zero for all practical purposes. It means that anyone looking for a job in 2009 could have found one in Iowa City in a short time, but finding a job in Detroit could have taken years. These staggering geographical differences are not just specific to periods of recession. Even in more normal times, unemployment in Detroit can be double the rate in fast-growing cities. And yet, unemployed people in Detroit do not all leave at the same rate. While college graduates are streaming out of that city, the flow of high school graduates is much slower, and the flow of high school dropouts is a mere trickle.

In total, almost half of college graduates move out of their birth states by age thirty. Only 27 percent of high school graduates and 17 percent of high school dropouts do so. These differences in mobility rates reflect the fact that some attend college out of state, but they mostly reflect differing propensities to look for work elsewhere. Using data from millions of individual histories from the economic census, the Notre Dame economist Abigail Wozniak matched workers in their late twenties to the economic conditions they faced in their state when those workers were eighteen and about to enter the labor market. Some of these young workers were fortunate and entered the labor market in states that, at the time, had a strong economy; others were less fortunate and entered the labor market in states with a weak economy. While being fortunate or unfortunate had little to do with schooling, how these young workers reacted to their fortune largely depended on their education. Wozniak found that among those who entered the labor market in bad times, a large portion of the college graduates relocated to states with stronger economies, while the majority of the high school graduates and high school dropouts did not move.

This implies that the job market for professional positions is a national one, while the job market for manual or unskilled positions tends to be more localized, so that people ignore good job opportunities in other cities. This is not just an American phenomenon but almost universal among rich countries. In the United Kingdom, the unemployment rates of highly educated workers in different regions are similar, because the high propensity to migrate tends to equalize job opportunities across regions, while the unemployment rate of less educated workers is vastly different. When Europeans are asked by pollsters whether they are "attached to their town or village," the number answering that they are "Not at all attached" or "Not very attached" is high in countries such as Finland, Denmark, and the Netherlands, which have high average educational levels, and low in countries such as Portugal and Spain, which have low average levels of education.

Reducing Unemployment with Relocation Vouchers

The relative lack of mobility of less educated Americans has large economic costs. We have seen that the changes in the global and national economy are causing an increase in inequality among workers with different skill levels, with the less skilled being hit the hardest. Differences in geographical mobility, coupled with increasing polarization among American cities, only exacerbate the problem. Thus, some of the earning inequality between highly skilled and low-skilled workers reflects mobility differences: if the less educated people were more able and willing to move to cities with better job opportunities, the gap between college graduates and high school graduates would shrink.

By being less mobile, less educated workers are also significantly more likely to be unemployed. Figure 10 shows the difference in unemployment rates for different educational groups over the past twenty years. Unemployment among all groups fluctuates, depending on the strength of the national economy. It was high in the early 1990s, reached its lowest level at the peak of the dot-com boom in 2000, and climbed up sharply during the Great Recession of 2008–2010.

But the most interesting feature of the graph is that in both good and bad years, college graduates—the group with the highest mobility—have the lowest unemployment rate, while high school dropouts—the group with the lowest mobility—consistently have the highest unemployment rate. High school graduates and workers with a community college education are in between. While the difference in unemployment rates reflects many factors, the willingness to move is an important difference among the four groups. It is not just that less educated individuals are more likely to be out of work at any particular time; they also have to deal with the long-term consequences. Evidence indicates that workers' skills tend to deteriorate during long bouts of unemployment, and this further widens the gulf between the skilled and the unskilled.

Why does a lack of education lead to lower mobility? For some, it reflects a dearth of information about opportunities elsewhere, a shortage of the kinds of skills necessary to make a big life change, and especially a lack of cash. Relocating is like an investment: you spend money up front, to cover the direct costs of the move and your living expenses until a job becomes available, in exchange for a better job later. But many unemployed workers with low skills

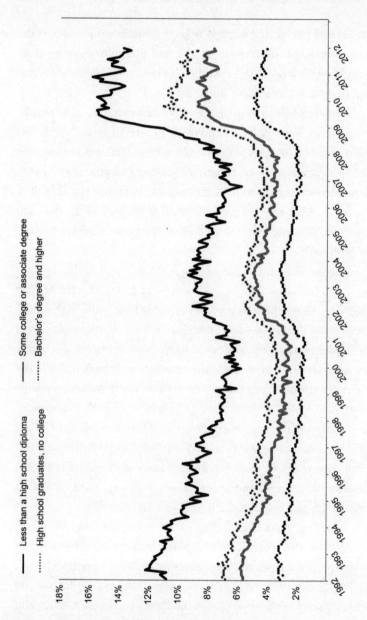

Figure 10. Unemployment rates by level of education

Source: Adapted from a graph by Bill McBride, calculatedriskblog.com.

Legend:
— Less than a high school diploma
······ High school graduates, no college
— Some college or associate degree
······ Bachelor's degree and higher

are unable to make this investment, because they have limited savings and limited access to credit. In this case, the lack of mobility is not a choice but the result of external constraints that limit people's freedom of movement. In other cases, the lower propensity to move reflects cultural differences between the two groups. Like some Italians, some less educated Americans choose not to move away, presumably because they value staying near their family and friends more than better job prospects. Although this has an economic cost, it is a perfectly legitimate choice.

This distinction between causes is important, because it suggests a policy reform that could end up helping those workers whose lack of mobility is not a choice. The unemployment insurance system, which was introduced during the 1930s, is essentially the same now as it was then. Currently, an out-of-work person who qualifies for unemployment insurance receives a check from the government that covers part of his previous salary. What is striking about the system is that it does not provide any incentive for unemployed workers to look for jobs in places with better labor markets. If anything, it discourages mobility from high-unemployment areas to low-unemployment ones, because it does not compensate for the difference in cost of living. If you are living off an unemployment check in Flint, you do not have a lot of incentives to move to Austin to look for a new job, because your housing expenses would double but your check would still reflect the cost of living in Flint.

The unemployment insurance system should be adjusted to reflect the vast and growing differences in economic fortunes among American cities. Unemployed people living in areas with above-average unemployment rates should receive part of their unemployment insurance check in the form of a mobility voucher that would cover some of the costs of moving to a different area. In other words, instead of encouraging out-of-work residents to remain in

Flint, the federal government could help them relocate to Texas (or wherever they might choose to go) with financial support that covers a portion of their moving expenses. This would help those who would like to move but are stuck because they lack cash.

Remarkably, this policy would also help those who are not willing to move. The reason is simple, although not widely recognized. If there are one thousand unemployed workers looking for jobs in a city where there are only one hundred job openings, the probability of each worker finding a job is one in ten. But if five hundred of these unemployed workers are encouraged to relocate by a mobility voucher, the probability that each of the remaining workers finds a job is doubled. In a climate of high unemployment, the fewer people like you who are looking for a job, the better your chances of finding one. This points to a surprising conclusion: unemployed workers who stay in a local labor market with high unemployment effectively impose a cost, or negative externality, on everyone else in that market, while workers who move away generate a positive externality. A mobility voucher is a way to deal with this. By increasing the number of workers who are willing to relocate, the voucher benefits both those who move, who end up with better jobs elsewhere, and those who stay, who end up with a better chance of finding a job. (Of course, this only works for the nation as a whole if the externality created by an unemployed worker is larger in cities with high unemployment than in cities with low unemployment—a reasonable assumption. Otherwise, mobility vouchers would simply shift the problem around, with no real benefit for total unemployment.)

In practice, a mobility voucher could take the form of an additional payment over and above the current unemployment insurance payment for those who move out of areas with above-average unemployment. Or it could come out of the current unemployment insurance payment, in the form of lower benefits for those who stay (with exceptions for those with health conditions or family constraints). The first case is a subsidy for those who move; the second is a tax on those who stay. If people cannot move because they don't have savings and/or they have limited access to credit, the effect on their mobility is likely to be larger with the first type of voucher. A combination of the two approaches is also possible.

This idea is not completely new. The government already provides a limited relocation allowance as part of Trade Adjustment Assistance, an obscure federal aid program that helps workers who have lost their jobs as a result of foreign trade. It is time to extend the allowance to include all workers receiving unemployment insurance.

In 1968 the Harvard economist John F. Kain proposed the theory of "spatial mismatch." Poor people and minorities, he argued, face a structural disadvantage in the labor market because of the geographical mismatch between the location of housing and the location of jobs within each city. His basic idea was that the poor tend to be concentrated in the urban core of American cities, far from many suitable jobs. Not only do the poor face higher commuting costs, but because they live farther from potential employers, they also have less information about job openings, and this further depresses their employment opportunities. Kain argued that this geographical dislocation, combined with lower car ownership rates and lack of efficient public transit, results in higher unemployment rates. The sociologist William Julius Wilson embraced the notion of spatial mismatch in his influential book *The Truly Disadvantaged*, which highlighted the role of mismatch as one of the root causes of racial differences in the United States. Effectively, the spatial mismatch theory attributes economic inequality in part to the patterns of residential segregation within each city.

While historically such differences may have played a role, today it is differences *across* cities that are more likely to be the source of mismatch. The divide between well-educated workers with high-paying, secure jobs and less educated workers with low-paying jobs is connected to the geographical divide between thriving cities and struggling ones. In the debate over inequality in America, people often overlook this aspect. As the gulf between the labor markets in American cities grows, the lower propensity of less skilled workers to move becomes more and more costly.

While the high mobility of well-educated Americans tends to be good for their careers, it presents state governments with a big challenge. By funding local universities and colleges, states heavily subsidize the higher education of their residents in the hope of fostering economic growth at the local level. In the United States, the current subsidy to students at public universities is on the order of 80 percent. As we saw, the overall level of human capital in an area is one of the most important drivers of local prosperity. State legislatures' support for higher education is based on the hope that it will raise labor productivity and attract innovative businesses. However, the fact that college-educated Americans are so mobile makes the states' efforts less effective.

A team of University of Michigan economists led by John Bound has found that the number of degrees conferred by local colleges and universities has only a modest effect on the number of university-educated workers within the state. States like Michigan and Ohio, with world-class systems of public higher education, struggle to retain many of their college graduates, who are more drawn to opportunities in California and New York. For the average recipients of bachelor's degrees, Bound and his coauthors found only a weak link between the number of students who graduated from a state university and the number who ended up staying in that state. For recipients of M.D. degrees, they found no connection whatsoever: the number of doctors who stayed in Michigan had nothing to do with the number of doctors produced by the University of Michigan. Because of the high mobility of college graduates, the Michigan economists concluded that states have limited power to influence the skill levels of their workforces in a meaningful way by investing in higher education. The pull of innovation hubs dwarfs their efforts. This is great news for the cities that attract the college graduates—these cities effectively receive free human capital paid for by someone else. But it significantly limits the ability of struggling states to build a sustainable base by investing in higher education. Bound's finding also has interesting implications for educational policy. It suggests that the financing of public colleges and universities should not be left solely to states. Given that the social benefit of investment in higher education is not contained within the borders of a state, an efficient educational policy is one in which the federal government plays a role in supporting part of this investment.

We now turn to another important aspect of Americans' mobility: its relation to real estate prices. We have seen that there are large and growing differences in wages among American cities. One of the primary reasons that people aren't moving en masse to San Francisco or Boston, despite the promise of higher wages, is that these cities are very expensive to live in. How is cost of living affecting the Great Divergence?

The Surprising Relationship Between Inequality and Real Estate

The city of Norilsk, in northern Siberia, sits on one of the earth's largest reserves of nickel and platinum. Because nickel is a necessary component of steel, Soviet planners made the development of Norilsk a priority during the 1930s. Stalin sent a delegation of experts to explore the region, and the experts reported that it would be difficult to attract workers to the city even by offering premium wages. Conditions there were simply too hellish: extremely cold temperatures, sometimes dropping to 45 degrees below zero, five months of winter darkness, and a depressing landscape with virtually no vegetation made it one of the most hostile environments on the planet. Although the precious metal reserves in the area could, in theory, support many jobs, planners could not come up with wages high enough to compensate workers for the horrible living conditions. For someone like Stalin, of course, these were trivial details. The feared state police, the NKVD, took over responsibility for the development of the city and turned it into a gulag—a Soviet labor camp. About 100,000 political prisoners died building the city and laboring in its mines. For decades the melting of the snowpack during the summer months revealed the bones of workers who had perished.

In the Soviet Union, as in other communist regimes in Eastern Europe and China, the state had the power to forcibly move labor where it was needed. This gave rise to "artificial" cities such as Norilsk—cities that would not have existed in a free society. In the United States, workers are free to choose where they want to live. As we saw, Americans take advantage of this and move around more than citizens of most other countries. But there is a catch to this freedom. Living in places that are perceived as more desirable, either because they offer a higher quality of life or because they offer better jobs, tends to cost more. This is not surprising. Unlike the Soviet economy, which assigned resources based on a five-year plan, a market economy uses prices to allocate resources, and in this case the scarce resource is land in attractive cities. If a city has great weather, Americans tend to move there in large numbers, and in doing so they bid up the price of real estate. Good weather may not have a sticker price, but we implicitly pay for it, just as we pay for a nicer car or a larger TV. The same is true for good public schools, low crime rates, and excellent local restaurants. Every attractive feature of a city ends up being capitalized, at least in part, into higher property values.

This simple observation has an unexpected implication: those who actually end up benefiting from these features are not necessarily those who are directly affected by them. Pollution levels in Southern California have decreased dramatically over the past twenty years, especially in Los Angeles, thanks to cleaner-burning gasoline and more aggressive regulation. Some neighborhoods have experienced more improvements than others, with ozone reductions ranging from 3 to 33 percent, depending on the area. You might think that residents of the neighborhoods that have experienced the largest pollution declines are the net winners, but that depends on whether they own or rent their homes. One study found that the larger the decline in pollution levels, the larger the increase in the desirability of the neighborhood and therefore the higher the price of real estate. In one low-income neighborhood, for example, ozone concentrations declined by 24 percent but housing costs increased by 10.8 percent. The price increase benefited property owners, who became both healthier and richer, but left renters healthier but poorer. Effectively, the price change acted as an unintended redistribution mechanism that shifted some of the benefits of air-quality improvements away from one group and toward another.

The same principle applies when a city labor market improves and local jobs are created. In the United States, we can see a clear correlation between local labor market conditions and the cost of living. Table 3 shows the metropolitan areas with the highest and lowest costs of living today. To create the table, I used data on about one million households, including both renters and home-owners, and data from the Bureau of Labor Statistics on the price of consumer goods. To measure cost of living, one needs to add up the local price of all the things consumed by residents. How does the average American household spend its money? Most people give the wrong answer. People tend to grossly overestimate the amount of money they spend on food, gas, and groceries, probably because they purchase these items regularly. In reality, the average American spends only 14 percent of her income on food and bev-erages and 17 percent on transportation. This is not very much. The other categories account for even less of the family budget: apparel (3 percent), medical care (6 percent), recreation (5 per-cent), education and communication (6 percent). (The way Ameri-cans divide their family budget is fairly similar to the way families in other countries do, with the main exception being Italian fami-lies, whose share of clothing expenditures is double that of Ameri-cans.) By far the largest item in the budget is housing, which ac-counts for 40 percent of spending. This means that most of the differences in cost of living among metropolitan areas reflect dif-ferences in the cost of housing, which in turn mostly reflect dif-ferences in the cost of land. Other differences arise from the price of local services—things like haircuts and restaurant meals—but these count considerably less, because their share of the budget is smaller. Moreover, they too mostly reflect the cost of land. For example, a haircut is more expensive in New York than in Dallas because it costs more to rent a store and because the salary of the hairstylist is higher to compensate for the higher cost of living. The same is true for restaurant meals, therapy sessions, legal ser-vices, and nanny services.

The table confirms that areas at the top of the list tend to be the ones with the strongest labor markets—the ones where wages and productivity are highest. San Jose is first, followed by Stamford

TABLE 3: METROPOLITAN AREAS WITH HIGH AND LOW COSTS OF LIVING

HIGHEST COST OF LIVING	LOWEST COST OF LIVING
1. San Jose, CA	271. Youngstown-Warren, OH/PA
2. Stamford, CT	272. Lima, OH
3. San Francisco–Oakland–Vallejo, CA	273. Terre Haute, IN
4. Santa Cruz, CA	274. Sharon, PA
5. Santa Barbara–Santa Maria–Lompoc, CA	275. St. Joseph, MO
6. Ventura–Oxnard–Simi Valley, CA	276. Lynchburg, VA
7. Boston, MA	277. Williamsport, PA
8. Honolulu, HI	278. Joplin, MO
9. Santa Rosa–Petaluma, CA	279. Brownsville–Harlingen–San Benito, TX
10. Salinas–Seaside–Monterey, CA	280. Duluth-Superior, MN/WI
11. New York–Northeastern NJ	281. Johnson City–Kingsport–Bristol, TN/VA
12. Washington, DC/MD/VA	282. Altoona, PA
13. Los Angeles–Long Beach, CA	283. Alexandria, LA
14. San Diego, CA	284. McAllen-Edinburg-Pharr-Mission, TX
15. Seattle-Everett, WA	285. Danville, VA
16. Trenton, NJ	286. Gadsden, AL
17. Bridgeport, CT	287. Anniston, AL
18. Fort Lauderdale–Hollywood–Pompano Beach, FL	288. Johnstown, PA
19. Austin, TX	
20. Anchorage, AK	

and San Francisco. Many American innovation hubs are in the top group—Boston, Washington, D.C., San Diego, Seattle, and Austin. Anchorage is an exception, because many of its necessities have to be imported. Because the data reflect the entire metropolitan area, New York is only number eleven; taken alone, the city of New York would be at the top of the list. By contrast, the areas with the most affordable cost of living tend to have the weakest labor markets. At the very bottom of the list we find Johnstown, Pennsylvania, a declining manufacturing town, where the cost of living is four times lower than that of San Jose. Other metro areas near the bottom include Anniston, Alabama; Gadsden, Alabama; and Danville, Virginia. The relation between the strength of a labor market and the cost of housing is not deterministic, but it depends on several factors, including quality of life (better quality of life means higher housing costs, all other things being equal) and how easy it is to build new houses to accommodate increases in demand (easier housing development means lower costs).

These facts have a bearing on how we interpret measures of inequality among workers and between cities. Let's start with the latter. When the labor market in a city strengthens, both workers' earnings and the cost of housing tend to increase. These increases have two separate effects on residents. First, the increase in housing costs offsets some of the increase in salaries. In cities like Johnstown, people have low nominal salaries, but since mortgages are lower than elsewhere, an average salary has more purchasing power. By contrast, people in New York, Washington, and Boston have higher nominal salaries, but their effective salaries are not as high, because much of their wage tends to go toward paying the mortgage. This helps explain why not everyone has left Johnstown to move to Boston or New York. In practice, differences in average earnings among U.S. cities adjusted for cost of living are about 25 percent smaller than unadjusted differences.*

However, this is not the end of the story. Just as with improvements in air quality, the effect of a strong labor market on a family ultimately depends on whether that family belongs to the 70 percent of Americans who own their homes or the 30 percent who rent. Homeowners in strengthening labor markets gain twice, both because of higher wages and because of higher property values. For them, the effect on well-being is larger than the increase in purchasing power because of the capital gains on their property. This highlights an unexpected conclusion: a significant part of the wealth created by America's dynamic innovation sector accrues not just through the labor market but through the housing market. These capital gains are an important channel through which residents of innovation hubs benefit from the strength of their local economy. For renters, however, the effect of higher earnings is tempered by the increase in their monthly housing costs. Therefore, the ultimate effect on their well-being depends on which of these two forces prevail. The larger the increase in wages and the smaller the increase in rents, the better for them. As in the case of air quality, the change in real estate prices effectively redistributes the wealth created by job growth from one group to another. As we will soon discover, local governments have the power to manage the increases in local cost of living and can therefore determine

---

* If the whole population were perfectly mobile and all American cities offered the same quality of life, purchasing power would be perfectly equalized among all cities. However, cities vary in the quality of life they offer, and not every American is willing to relocate to a place that offers higher wages, because people often prefer to live in one place rather than another. Thus in practice purchasing power and well-being are not perfectly equalized across the United States. (In economic jargon, well-being is equalized for marginal individuals but not for inframarginal ones. See Moretti, "Local Labor Markets," for a more detailed discussion.)

whether homeowners or renters are the ones to gain the most from a strengthening labor market.

This relationship between local labor markets and cost of living also affects the way we think about inequality between workers. Most of the public debate on inequality focuses on the striking differences in salaries and incomes, but what really matters is how much people can buy with their earnings. When economists started measuring inequality this way, they found that the difference in consumption between rich and poor—from groceries to clothes, electronics to health care—is not as large as the difference in salary. How can the consumption gap between the rich and the poor be smaller than the income gap?

An important explanation for this apparent contradiction has to do with where people live. In recent research, I found that since 1980, the amount that the typical college graduate spends on housing has grown much faster than the amount the typical high school graduate spends. This trend does not just reflect better or larger houses owned by college graduates. It mostly reflects differences in where groups with different skills tend to congregate. As we have seen, over the past three decades, jobs for college graduates have increasingly concentrated in expensive metropolitan areas—brain hubs like San Jose, San Francisco, Boston, New York, and Washington, D.C.—while jobs for high school graduates have increasingly concentrated in heartland cities with a low cost of living. While in 1980 the difference in housing costs between the two groups was small, it has grown by more than three times. This is important, because it implies that college graduates end up spending more for housing and therefore have less money for other goods and services. It is as if college graduates have experienced a higher inflation rate than high school graduates. Therefore the difference in living standards between highly educated Americans and less educated Americans, while large, is actually somewhat smaller than you might think.

Gentrification and Its Discontents

Just as improvements in air quality can have unintended consequences, a stronger labor market can sometimes have a dark side. Higher real estate prices can displace the poor, significantly altering the mix of residents in a community. Eventually these changes can affect a city's very identity. For example, think of Boston in the 1970s. Its economy was in terrible condition, bogged down by an old manufacturing base and high unemployment. But over the past three decades it has flourished, thanks to jobs in innovation and finance. The transformation was not just economic but also demographic and cultural. It resulted in a profound remaking of the social texture of the city, its urban form, and its quality of life. While many of these changes were for the better, there were also significant social costs. Many longtime residents ended up being priced out of their own neighborhoods. And even those who stayed were not spared, as the character of some communities shifted in sudden and sometimes uncomfortable ways. Those who moved to Boston between 1990 and 2010 tended to have college degrees and professional occupations. Those who moved out tended to have low levels of schooling and nonprofessional jobs. The lifestyles, values, and social identities of the two groups could not have been more different.

The debate about the costs of local economic development can get acrimonious. Local activists in cities such as Cambridge, Berkeley, Washington, D.C., and Santa Monica love to hate this sort of economic change, arguing that it should be stopped at any cost because it ends up hurting communities. While it is clear that there are costs, it is useful to clarify who bears them and what the best way to minimize them is.

As we have seen, original homeowners benefit from gentrification. It is important to recognize that this group can be socially quite distinct from the gentrifiers—the college-educated professionals, the innovators, the entrepreneurs. Almost by definition, a gentrifying neighborhood is one in which many of the original residents, including the property owners, are not particularly wealthy. Take the Mission District, the neighborhood of San Francisco where I live. It is one of the areas of the city that has been most affected by the influx of college-educated high-tech professionals. Since it is close to the freeway, many workers in Silicon Valley who prefer an urban lifestyle end up here. Remarkably, the people who are benefiting most from this influx of high-tech workers are the largely Latino homeowners who have been selling their property to the newcomers—people like the Mexican American couple who owned a nice two-story Victorian near my house that had been in the family for decades. They decided to sell it for $950,000 and move to the suburbs, where they could buy a similar-sized house for half the price and live off the balance.

What about all the other residents, people who never had any property to begin with? In many large urban areas, most residents are renters and are therefore hit hard by an increase in cost of living. This is especially painful for the elderly and those with low incomes who end up relinquishing their houses, memories, neighbors, and social networks—in a word, much of their lives—and have to start from scratch somewhere else. What should we do to protect them from sudden displacement?

The typical reaction in many communities is stringent land-use regulation designed to slow down socioeconomic change. These laws tend to come in two flavors. The first focuses on commercial real estate and seeks to moderate gentrification by limiting the number of new office buildings. One of the most extreme cases is the city of Berkeley, which in an effort to protect "good blue-collar jobs" has effectively stunted high-tech growth in the entire west side of the city. Large parts of eastern San Francisco are also slated for light manufacturing in the vain hope that the industry will bounce back. The second kind focuses on residential real estate and seeks to limit new market-rate construction, particularly in transforming neighborhoods. In effect, the first kind seeks to limit the inflow of new employers in the innovation sector, while the second seeks to limit the inflow of new residents. Both aim to reduce private investment with the intent of preserving the existing economic and cultural demographics.

In my view, both approaches are misguided and unlikely to be effective at managing gentrification. Constraining new high-tech office buildings amounts to reducing the number of jobs that a city can create, because it is quite unlikely that factories will open in the urban core of cities like San Francisco and Santa Monica. Because of the multiplier and spillover effects, this policy ends up hurting the very people it is intended to help. Indeed, the most important lesson of the multiplier and spillover effects is that unskilled workers in a city have much to gain from the fortunes of the more skilled workers who live next door, as their very livelihoods often depend on sustained growth in the innovation sector: more skilled residents in a city mean more and better jobs for the less skilled. The policy also hurts nonresidents. As we have seen, innovation hubs are among the most productive areas in the United States, and this higher productivity attracts workers from everywhere in the nation. This level of productivity cannot be easily replicated elsewhere, because of the strong forces of agglomeration. Thus, curtailing job creation in America's innovation hubs is apt to result in a net job loss for the country. It is a terrible waste of resources—one that makes our unemployment rate worse.

Curtailing new residential developments also makes little sense. It is the equivalent of creating jobs in a city but then denying those jobs to any applicant who comes from somewhere else. Moreover, it is likely to accelerate the displacement of poor residents, not to slow it down. The reason is quite simple: rationing new housing in a city invariably results in even higher real estate prices. It makes intuitive sense: if there is high demand for housing in a city, reducing supply can only raise the price. In a series of recent studies, the urban economist Ed Glaeser and various collaborators have uncovered clear evidence that cities that adopt more restrictive residential development policies invariably end up with higher housing costs relative to wage levels. By contrast, cities that are proactive in allowing urban housing development end up with lower housing costs.

The real solution to the problem of gentrification is exactly the opposite of restricting new residential development. Instead of limiting new housing, innovation hubs should encourage it. If managed correctly through smart growth policies, more housing does not mean more sprawl and congestion, especially if it is concentrated in the urban core and is accompanied by an expansion of the public transit system. These kinds of progressive urban development policies can significantly mitigate the negative effects of gentrification while promoting the serendipitous urban social interactions that foster knowledge spillovers and innovation.

A good example is Seattle. When economic conditions there began to improve, thanks to the expansion of high-tech jobs, the city decided to increase the number of new housing units available to families by allowing a significant amount of infill urban development—the type of development that focuses on making an area denser by renovating existing buildings and developing empty lots, thus avoiding sprawl. This increase in supply kept real estate prices in check. While there clearly were price increases, they were lower than those of cities like San Francisco and Boston, which aggressively limit new housing. Essentially, this acted like a redistribution mechanism that favored renters over homeowners. It meant that a larger share of the wealth created by the rise of the local high-tech sector went to the former group instead of the latter.

Seattle was also fortunate to have farsighted business leaders. In contrast to those in most other American cities, major retailers in Seattle decided to stay downtown. The Nordstrom family, together with the other department stores at the time — most important, Frederick & Nelson — wanted to head off the "exodus to the mall" phenomenon. This decision encouraged urban planners to embrace a retail-based urban center, an unusual move for American cities. Things would be very different today if Nordstrom had fled. The simultaneous growth of high-tech jobs in and around the downtown area and high-density housing in the city's walkable neighborhoods reversed middle-class flight from the city center and ultimately resulted in a lower crime rate, vibrant cultural offerings, and new restaurants. As new residents flocked to the urban core, public schools experienced noticeable improvements. Test scores increased, and not just for those children who had well-educated parents but also for children whose parents were less well educated and who had non-high-tech jobs.

In the end, from the point of view of a city, gentrification is a good problem to have, because it is a sign of economic success and job growth. Dozens of decaying cities would love to have this problem. At the same time, gentrification has serious social consequences. The solution is not to discourage local job creation in the innovation sector, hoping that manufacturing jobs will magically return. The solution is to manage the process of economic growth in smart ways, to minimize the negative consequences for the weakest residents and maximize the economic benefits for all.

6

◆

POVERTY TRAPS AND SEXY CITIES

W E LIVE IN A COUNTRY that is rapidly growing apart. Thriving industries tend to cluster in some cities but not in others. These cities create good jobs and generate good salaries while others lag further and further behind. People can move from failing cities to prosperous ones, but as we have seen, this is not a panacea. The question therefore is how to help communities that are stuck with the wrong mix of jobs and skills. Can we help cities like Flint, Mobile, and Visalia create a self-sustaining local ecosystem that creates good jobs in the community?

This is a challenging question to answer, but one way is to look at how existing innovation clusters were created and see whether that process can be reproduced elsewhere. The history of the biotech industry is particularly enlightening. In the spring of 1973, Herbert Boyer and Stanley Cohen invented a recombinant DNA technique that changed the course of life science research forever.

Almost immediately, dozens of private biotech labs appeared all over the United States—among other places, in Houston, Long Island, Cincinnati, Montgomery, Cambridge, Philadelphia, northern New Jersey, Miami, Palo Alto, Emeryville, Los Angeles, and La Jolla.

Today the three locations with the largest concentration of private biotech firms are the Boston-Cambridge metropolitan area, the San Francisco Bay Area, and San Diego. Their share of industry jobs keeps increasing. Remarking on the Bay Area cluster, the biotech venture capitalist Dr. Charles Homcy, a partner of Third Rock Ventures, said, "Finding great science and great scientists and the next innovative platform to revolutionize medicine—there are few places to do that. This is one of them." But in 1973 it was not so obvious where the industry was going to cluster. There was nothing to suggest that the cities now at the top of the chart were necessarily going to be the winners.

Of the three, San Diego, a quiet community that attracted mainly retired navy sailors, fishermen, and tourists, was the most unlikely location for a biotech cluster. Describing the early days of the industry, the Stanford sociologist Walter Powell reports that in the 1980s, "Torrey Pines Road in La Jolla, now the epicenter of 'biotech beach' in San Diego County, was more widely known for its golf courses and gorgeous beaches than for its laboratories." Even Cambridge was not a slam dunk. Kendall Square in Cambridge is now full of space-age biotech labs, but even as late as 1985 it was "riddled with decaying textile factories." The liberal academic establishment in Cambridge was at first hostile to the industry, mainly because of its opposition to genetic engineering. The biotech pioneer Biogen learned this the hard way. In reconstructing the history of the cluster, Powell noted that "public uproar over 'Frankenstein factories' led the founders of Biogen to incorporate initially in Switzerland to avoid the controversies in Cambridge, and co-founder and Nobel laureate Walter Gilbert had to take a leave of absence from Harvard University."

So why did biotech put down its roots in these three locations? The conventional wisdom is that all three had premier universities: Cambridge has Harvard and MIT, the Bay Area has Stanford, Berkeley, and the University of California at San Francisco, San Diego has the University of California at San Diego. On a superficial level, this answer makes sense: academic research is crucial to biotech firms, with their emphasis on basic science. Therefore, we might expect that proximity to academic institutions played a fundamental role in biotech firms' location decisions.

But if we look a little deeper, we realize that this isn't the whole story. It is a classic case of after-the-fact rationalization. In the United States, there are 1,764 four-year colleges and 662 universities. The average metropolitan area has five colleges and two universities. It would be difficult for a biotech cluster to sprout without being physically close to a university. Even if we were throwing high-tech clusters randomly on a map of the United States, they would probably land only steps from a university.

Is it a question of being close to a *top* university, or a top biology department in a top university? This is not the answer either. When biotech appeared in the 1970s, at least twenty excellent research universities with world-class biology departments or research hospitals existed in cities as diverse as New Haven, New York, Philadelphia, Baltimore, Atlanta, Chicago, Madison, Denver, Cleveland, Houston, Pasadena, Ann Arbor, and Los Angeles. These were all attractive locations, but they did not all develop major clusters.

When the Stars Are Aligned

In 1998 the sociologist Lynne G. Zucker and the economist Michael Darby came up with a surprising theory. In a fascinating and now classic article and in a series of subsequent studies, they argued that what really explains the location and success of private biotech companies is the presence of academic stars—researchers who have published the most articles reporting specific gene-sequencing discoveries. Among top universities, some institutions happened to have on their faculty stars in the particular subfield of biology that matters for biotech; others had comparable research standing but did not have stars in that specific subfield. The former group created a local cluster of private biotech firms while the latter did not. The data suggest that the magnetic effect of academic stars is impressive. Zucker and Darby estimate that stars are more important than proximity to venture capital firms or the effect of government funding. It is not just that stars explain where and when biotech startups appear on the map; they also affect which startups survive and thrive and which ones stuggle and disappear.

As Zucker and Darby point out, success in high technology, especially in its formative years, comes down to a small number of extraordinary scientists with vision and a mastery of breakthrough technology. Indeed, we can't overestimate the impact that these unusual individuals have on the economic development of cities and regions. With $350 billion in total investment worldwide to date, almost four hundred biotech medicines, and one thousand experimental compounds currently in clinical trials, the biotech industry can bring thousands of good jobs and considerable prosperity to a community.

There are two reasons for the power of stars. First, scientists and researchers in private-sector startups need to be physically close to frontier academic research in order to remain on the cutting edge. Attending regular academic seminars, engaging in informal discussions, and hearing what others are working on and what progress they are making are critical to forming and developing new ideas. Employees of private-sector research firms can reap the benefits of these knowledge spillovers only when their labs are physically close to those of top academic researchers. A second reason is that stars are often personally involved in the creation of leading private-sector startups. Zucker and Darby find that the typical pattern is for an academic entrepreneur to help establish a firm in the gene-sequencing area that he has pioneered while he is still on the faculty at a university.

The moral of the story is that Cambridge, San Diego, and the Bay Area were lucky. Where the stars lived when biotech emerged in the mid-1970s was to some extent random: it could have been any of the 187 American cities with a university, or at least one of the 20 cities with a top biology department. But what happened later was not random: the self-reinforcing nature of clusters means that once a cluster has started, it keeps attracting companies and workers. First movers benefit from this lock-in effect, and early advantages become magnified over time. The attractive nature of economic development ensures that even today the industry keeps agglomerating in Cambridge, San Diego, and the Bay Area. Although the impact of stars on the creation of startups is fading over time as the industry matures, their effect on the local economy is long-lasting.

What Biotech and Hollywood Have in Common

Biotechnology is not an isolated case. It is representative of how many innovation clusters get started. Throughout history, hubs of innovative activity have agglomerated in unlikely places. Consider another important industry whose success depends on stars: motion pictures. At the beginning of the twentieth century, film was the hot new thing, competing with theater to establish itself as a respectable medium of entertainment and facing formidable technological and managerial challenges common to all new industries. Everything had to be invented from scratch, from the shooting and editing of a film to its production and distribution.

In 1913, the year before World War I began, the movie industry was largely concentrated in New York, where the major studios and the biggest stars were, with smaller outposts in Chicago, Philadelphia, Jacksonville, Santa Barbara, and Los Angeles. By 1919, one year after the end of the war, 80 percent of American movies were made in California. Charlie Chaplin and countless other stars had moved west, and Los Angeles had three times as many motion picture establishments as New York. The golden age of Hollywood had begun. By the mid-1920s, Los Angeles had further consolidated its position as the world's premier location for film, and the term *Hollywood* no longer just referred to a quiet neighborhood west of downtown and had begun to be used as a generic term for the entire filmmaking world. This era of unprecedented artistic achievement and commercial success peaked in 1940, when Hollywood studios produced about four hundred films a year and 90 million Americans went to the movies each week. By then the economy, society, and culture of Los Angeles had changed forever. Movies had become big business, generating tens of thousands of local jobs, and were responsible for a significant portion of the city's prosperity.

The transformation of Los Angeles from a small, provincial outpost, remote from everyone and everything, into a cosmopolitan center of artistic creation is a truly breathtaking tale. Its dynamics track what we see happening in more modern innovation clusters. As more and more actors, studios, and specialized service providers (stage technicians, musicians, location scouts, costume designers, and so forth) congregated in Hollywood, the forces of agglomeration sustained an accelerating upward trajectory. This agglomeration made Los Angeles the place to be, and made it increasingly difficult for other locations to compete. In exactly the same way that Silicon Valley and Seattle today attract the best and the brightest Chinese and Indian engineers, Hollywood at the time became a magnet for talented immigrants, mostly European, many of them Jewish: great directors like Ernst Lubitsch, Alfred Hitchcock, Fritz Lang, and Michael Curtiz and great actors like Rudolph Valentino, Marlene Dietrich, and Ronald Colman.

Although the economic forces that are responsible for Los Angeles's rapid rise are clear, the initial seed is not. Why Los Angeles? The conventional explanation has always been that the motion picture industry needed to be based in Los Angeles because of its good weather; the cold New York winters presented technical challenges for outdoor filming. But while the weather was important, it could not have been the decisive factor. This is another case of after-the-fact rationalization. Los Angeles is not the only city in America with a good climate. And Berlin, London, Paris, and Moscow—none of them with mild winters—all remained film capitals.

In 2006, the UCLA geographer Allen Scott proposed a much better explanation. He pointed out that Los Angeles's rise hinges on the year 1915, when a powerful combination of commercial and cultural forces transformed the city. The event that precipitated Hollywood's ascent had to do with a real star, the pioneering director D. W. Griffith. The inventor of a number of new techniques that would define filmmaking for decades, including the close-up, the flashback, and the fade-out, Griffith became so influential that Charlie Chaplin called him "the teacher of us all." The key moment for Hollywood occurred in 1915, when Griffith shot the first big-budget blockbuster in history, *The Birth of a Nation*. With a production cost of $85,000—five times more than that of any film made before—*The Birth of a Nation* earned over $18 million in sales, far more than any other film of the silent era. It was the film that brought motion pictures firmly into the mainstream and made them appealing to middle-class audiences, who until that point had considered film inferior to theater. In the process it planted the seed of Los Angeles's future success. Three years after the film was made, the city already had twice as many workers in the film industry as New York, and the gap kept growing every year for the next two decades. The process of agglomeration had started, and there was no turning back.

With the benefit of hindsight, the location of industries appears inevitable. Today we immediately associate Los Angeles with movies, New York with finance, Silicon Valley with computers, Seattle with software, and the Raleigh-Durham area with medical research. But this is not how people saw it before these industries settled in their respective cities. In 1910 there was little in Los Angeles that suggested it was going to become the film capital of the world. There was nothing in the Raleigh-Durham region in the 1960s that indicated it would become a biomedical research capital. In the 1970s, Seattle seemed like the last place that would become a global hub for software development. Cambridge, San Diego, and San Francisco happened to have the right kind of stars at the right time. By contrast, the location of traditional manufacturing is easier to explain, because it can often be traced to such physical factors as access to a harbor or proximity to natural resources. There is a reason that Chicago, Detroit, Toledo, Buffalo, and Cleveland grew into sprawling manufacturing clusters in the nineteenth and twentieth centuries, and it has to do with cheap transportation of heavy materials over waterways.

The history of high-tech clusters indicates that while we understand fairly well what happens after clusters are established, we often have a hard time predicting them. We have an even harder time creating them. Even Silicon Valley, arguably the most important cluster in the United States, hardly appears planned. Military research had a lot to do with its beginning, but the cluster in the Valley did not take root because military officials sat down and decided to create an innovation hub in the region. In 1940 the peninsula south of San Francisco was a quiet agricultural region with a comparative advantage in fruit production. The arrival among the orchards of William Shockley, the legendary high-tech pioneer who invented the transistor, was the seed that sparked growth of the local innovation industry. When some of Shockley's disciples created the first integrated circuit at Fairchild Semiconductor, it became clear that the seed had germinated: the process of clustering had begun. That serendipitous seedling was the starting point of an economic miracle that eventually brought millions of jobs to the region.

While it is true that Shockley was connected to Stanford—a fact that most histories of the Valley point to as proof that the Valley owes its existence to the university—at the time Stanford was just one among the many universities in America, and not even the best one. Of course Stanford did play a role, but it was less deterministic than many people think. A research university was necessary but far from sufficient for the birth and coming-of-age of the Valley. If Shockley had decided to locate in, say, Providence, which was then an area with a significantly more developed industrial base than Palo Alto, Silicon Valley might today be clustered in Rhode Island, and we would be reading dozens of books on how Brown University caused the cluster.

Visionaries have been trying to build thriving cities from the time that people started living in them. Utopian communities have always ignited people's imaginations, with their promise of curing social ills through enlightened planning and strong values. In most cases these communities have not lasted. In 1928, Henry Ford tried to establish a new industrial center called Fordlandia, building it from scratch on virgin land. His vision was to apply the rational efficiency of Ford engineering to the building of an ideal community in the middle of the Brazilian rainforest to harvest rubber for Ford's tires. As it turned out, it was difficult to engineer utopia. Ford's experiment proved to be a disaster for residents and investors alike. It was sold at a big loss just seventeen years after it had been inaugurated with great fanfare.

Struggling communities all across America are now trying to reinvent themselves and attract good jobs. How should governments aid in this effort? Ever since Michael Porter popularized the catchy concept of cluster building in the early 1990s, cities and states have been trying to engineer clusters through a variety of public policy measures, which economists call *place-based policies*. They are effectively a form of welfare, but they target cities, not individuals. About $60 billion is spent annually by states and the federal government on these policies—more than is spent on unemployment compensation in a normal year. But the economic logic behind such measures is rarely discussed and even less frequently understood.

Do these policies work? To answer this question, we must examine the underlying ideas more closely and rigorously evaluate their economic rationale. We will discover that just as Henry Ford faced challenges in building a city from scratch, local governments face challenges in reorienting regional economics. Understanding when government intervention makes sense and when it doesn't is a crucial first step in setting sound policies.

Poor but Sexy

The economy of a successful city is based on a remarkable equilibrium between labor supply and demand: innovative companies (the labor demand) want to be there because they know they will find workers with the skills they need, and skilled workers (the labor supply) want to be there because they know they will find the jobs they are looking for. The economy of a struggling city is the opposite. Even if real estate is dirt cheap, skilled workers do not want to be there, because they know there are no jobs; innovative companies do not want to be there either, because they know there are no skilled workers. It would be in the interest of one group to move if the other did, but neither wants to go first. It is a Catch-22.

Broadly speaking, there are two approaches to revitalizing a struggling urban area. One—I will call it the *demand side approach*—tries to attract employers with the hope that workers will follow. This often involves providing incentives and tax breaks to make a place attractive to companies. The other, which I will call the *supply side approach*, tries to attract workers with the hope that employers will follow. It involves improving a city's local amenities to lure talented workers. In essence, the first strategy is about bribing businesses, while the second is about bribing people.

Ten years ago, the idea of revitalization through amenities suddenly became very fashionable. Richard Florida's influential books publicized the notion that the "creative class" is particularly sensitive to the quality of life and that local economic growth hinges on making a city interesting and exciting for its members. He writes, "Seattle was the home of Jimi Hendrix and later Nirvana and Pearl Jam as well as Microsoft and Amazon. Austin was home to Willie Nelson and its fabulous Sixth Street music scene before Michael Dell ever stepped into his now famous University of Texas fraternity house." To prosper, a city needs to have culture and liberal attitudes—in short, it needs to be cool. Florida's prescription became a popular remedy as scores of public officials and local policymakers embraced the idea that public redevelopment resources should be mainly focused on improving city amenities to attract the "creatives." Hundreds of millions of dollars were spent and are still being spent in communities across America, from Pittsburgh to Detroit, Cleveland to Mobile. In 2003 the state of Michigan launched an ambitious campaign and a slick website called "Cool Cities" in an effort to rebrand former industrial towns like Flint and Detroit as sexy and livable for the creative class. The Ford Foundation has made $100 million available to fund art spaces that "spur economic development in their surrounding areas." "The arts have become the default tool for economic development," Karrie Jacobs recently wrote in *Metropolis*. Florida's fame was further increased by his claim that a good predictor of a community's economic success was its openness to gays.

It is certainly true that cities that have built a solid economic base in the innovation sector are often lively, interesting, and culturally open-minded. However, it is important to distinguish cause from effect. The history of successful innovation clusters suggests that in many cases, cities became attractive because they succeeded in building a solid economic base, not vice versa. For example, people who visit Seattle today will find it culturally vibrant, with great restaurants and tolerant attitudes. They may be tempted to conclude that the innovation sector grew in Seattle because creative types wanted to live there. But as we discovered earlier, it was exactly the other way around. For all its Jimi Hendrix connections, Seattle was not particularly attractive in 1980. It was gritty and depressing. Its residents were deserting it by the thousands. It became a lively cosmopolitan playground for educated professionals *after* it started attracting all the high-tech jobs.

Cleveland has attractive cultural amenities, including a first-rate orchestra and art museum and a spruced-up downtown, but it has been unable to establish a new economic base. Although Austin, one of the fastest-growing innovation hubs in America, is quite pleasant, it's not as lovely as Santa Barbara, but even with its ideal weather, beautiful surroundings, and relaxed attitudes, Santa Barbara has a sleepy local economy virtually devoid of high-tech jobs. The United States is dotted with attractive cities—Miami, Santa Fe, New Orleans—that offer plenty of culture and tolerance but do not generate good jobs in the innovation sector. Italy offers a wonderful lifestyle but has one of the lowest penetrations of innovative industries among developed nations. Italy's problem is not the supply of creative talent—there is no shortage of smart, ambitious, college-educated young people—but the demand for creative talent. Millions of Italian youths are unemployed or underemployed, in large part because of the failure of the economic system to attract a robust innovation sector.

Of course there are exceptions, New York being the most important one. New York has always had a solid base of good professional jobs, but for thirty years the city was underperforming because of a high crime rate, low quality of life, and poor public services. Its economic renaissance can be traced to improvements in quality of life and the willingness of many skilled professionals to return to the city.

Today one of the world's coolest cities is Berlin. In the twenty-three years since the fall of the Wall, it has become a magnet for creative people from all over Europe. Thousands of young, college-educated Italians, Spaniards, and French men and women move there every year, attracted by its world-class cultural landscape, its countless galleries and incredible public art displays, its unparalleled mix of highbrow and alternative musical offerings, its edgy dance clubs that open no earlier than 1 A.M., its affordable gourmet restaurants, and its increasingly diverse ethnic food scene. Berlin's well-established progressive attitudes, gritty but interesting architecture, and tormented history inspire a feeling of experimentation that some say is reminiscent of New York in the 1980s. Berlin also has one of Europe's most affordable housing markets; government-subsidized, high-quality child care; good schools; and outstanding public infrastructure, including a massive new glass-and-steel central railway station and a brand-new airport. Strangely, the city also benefits from its past as a divided city: two zoos, three major opera houses, seven symphony orchestras, and scores of museums are the legacy of forty years of Cold War competition. Walking along the beautiful streets of the historic downtown, you cannot escape the impression that this unique mix of creativity and high quality of life is hard to surpass. It is perhaps not too surprising that over one million newcomers have moved to Berlin since reunification, many of them highly skilled.

There is only one problem with this picture: there are hardly any jobs. Over the past decade, Berlin had the highest unemployment rate in Germany—almost double the national average—and the second lowest growth in per capita income. While it is by far the most interesting and creative city in the country, one of Europe's capitals of cool, Berlin has so far failed to attract a solid economic base. Its openly gay, progressive mayor, who champions the city's bohemian persona, has famously defined his city as "poor but sexy."

Michael Burda is a Harvard-educated American economist who teaches at Humboldt University, Berlin's oldest and most prestigious university. Burda has been observing the local economy since he moved there in 1993, just four years after reunification. As we sat at one of Berlin's thousands of open-air cafés, sipping good wine and admiring the beauty of the city's slow-moving river and lively street life, Burda confirmed what the statistics say: tourism is one of the main sources of jobs. Some large German companies are based in Berlin—mainly because it's the capital—and there are some jobs in fashion and media. Recently a cluster of Internet startups has sprung up, but it is too small to make a difference in a region of almost 4 million people. By and large, the presence of global innovation companies is limited, income is lower than in the rest of Germany, and good private-sector jobs have lagged ever since the Wall came down. The reality is that Berlin survives mainly because history has made it a tourist magnet and because it receives enormous transfers of money from the rest of the country in the form of direct investment in new facilities and public-sector jobs. Richer states such as Bavaria and Baden-Württemberg have been subsidizing employment in Berlin for decades. This stands in contrast to Washington, D.C., which over the past twenty years has created a deep and self-sustaining high-tech cluster of private companies over and above its public institutions.

It is hard to think of a better place than Berlin to test the notion that innovation hubs can be grown simply by catering to the creative class. Richard Florida's argument is that increasing amenities for the creative class leads to an increase in the supply of labor, and this ultimately lifts a city's economy. But after twenty years of Berlin coolness, the supply of well-educated creatives vastly exceeds the demand. According to one study, 30 percent of social scientists and 40 percent of artists are jobless. Germany has a thriving high-tech sector and a thriving advanced manufacturing sector, but only a tiny part of those is located in Berlin. Only time will tell whether the city will turn into the next Silicon Valley, but for now that is not happening.

Glamour is not enough to support a local economy. Ultimately a city needs to attract jobs. This does not mean that quality of life does not matter. Neil Kumar, the vice president of engineering at San Francisco–based Yelp, told *USA Today*, "Our location has helped us attract a smart, cultured and diverse workforce," adding that the city itself is a key recruiting tool in the intense competition for engineering talent. "We're able to attract creative and tech talent because we are in the city," says Colleen McCreary, an executive at the social gaming giant Zynga, also based in San Francisco. "It's not the No. 1 factor, but when you put the whole picture together, it certainly comes into play." Competitors located in less sexy parts of Silicon Valley have reacted by offering free bus service—locally known as GoogleBuses—to take San Francisco residents to work. Google, Apple, Yahoo, and Genentech all have their fleet of biodiesel-powered buses equipped with Wi-Fi to take their employees to their peninsula headquarters. Employees can work, surf the Web, and sip free low-fat cappuccinos while being chauffeured to the office. Bikes and dogs are welcome onboard.

A good quality of life does help cities attract talent and grow economically, but on its own, it is unlikely to be the engine that turns a struggling community into an innovation cluster. If it is not working for Berlin, it is hard to see how it could work for Flint.

Can Universities Be an Engine of Growth?

Every city wants to have its own MIT. Planners, city officials, and local politicians are constantly scheming to re-create the remarkable innovation hub that has flourished around the most famous engineering school in the world. They have in mind innovators like Yet-Ming Chiang and Dharmesh Shah. Chiang is a fifty-three-year-old professor of materials science. In addition to teaching and writing academic articles with engaging titles like "Overpotential-Dependent Phase Transformation Pathways in Lithium Iron

Phosphate Battery Electrodes," Chiang is a serial entrepreneur. His academic research has resulted in the creation of four high-tech companies. The most prominent are A123 Systems, one of the world leaders in clean tech, and SpringLeaf Therapeutics, which designs wearable drug delivery systems. A123 Systems is often profiled in the media, because it uses nanoscale technology to create more efficient batteries for electric cars. Chiang is directly responsible for thousands of new local jobs—A123 alone employs 1,700 people—and indirectly responsible for thousands more through the multiplier effect. Besides being exceedingly smart, he is very successful at leveraging all the competitive advantages offered by the Boston-Cambridge innovation hub. "I try to find partners at MIT who are experts in areas where I am not, and leading entrepreneurs to work with on the business side," he told the press when he was recognized as one of Boston's top innovators in 2011. "Then we interact with the venture community, and hire the best young business and technical talents. I take maximum advantage of our ecosystem." Dharmesh Shah was a student at MIT when he founded HubSpot in his dorm. The company, an online marketing platform for small businesses, now employs two hundred people in downtown Cambridge.

But do universities really change a community's economy? The role played by universities in local development is complex. As we have seen, the number of college-educated workers is the key factor driving the economic success of cities. But college graduates are a very mobile group, and they do not necessarily stay in the city where they went to school unless market conditions are attractive. The majority of college graduates in New York were educated not at Columbia, New York University, or City University of New York but at schools in other cities or states.

My research shows that the presence of a college or university in a city increases both the supply of college graduates, by educating some and attracting others from outside, and the demand for college graduates, by making them more productive.* The demand effect comes through three channels. First, some businesses are created directly as a result of academic research, like the ones founded by Yet-Ming Chiang. A recent study indicates that the passing of the Bayh-Dole Act in 1980, which encouraged universities to exploit their innovations commercially, resulted in job growth for communities near universities. Since 1980, MIT has generated 3,673 patents; companies started by MIT graduates and faculty generate $2 trillion in sales each year. Stanford and Berkeley can make similar claims. While Larry Page and Sergey Brin were graduate students in Stanford's engineering department, they developed the technology underlying Google's fabled search engine for a research project. In its early days, Google even operated under the Stanford website, with the domain *google.stanford.edu*. Its first employee was a fellow student of Page's and Brin's who, thanks to that lucky break, is now a millionaire.

A second important benefit of universities is that academic re-

---

* Establishing the effect of universities is difficult. Presumably universities tend to appear in cities where there is a perceived need for one, so the presence of a university may simply be the effect, rather than the cause, of a skilled populace. To deal with the possibility of reverse causality, I used data on land-grant colleges, which were established in 1862, when Congress created the first major federal program to support higher education in the United States. Altogether, seventy-three land-grant colleges and universities were founded, with each state having at least one. I compared cities with no local university with cities that have a university today because they received a land-grant college in the nineteenth century. Because this was an act of the federal government and did not depend on local conditions, cities with a land-grant college are a better comparison group than all cities with a university. Cities with and without a land-grant college looked similar before receiving a land-grant college, but they are significantly different today. The presence of a land-grant college in a metropolitan area results in 25 percent more college graduates and significantly higher wages.

search generates the kind of knowledge spillovers discussed earlier, and this further fosters a local innovation sector. A study by Adam Jaffe has found that this spillover effect is particularly relevant in the areas of drugs, medical technology, electronics, optics, and nuclear technology. While some of the spillover accrues to companies everywhere, a significant part is local in scope.

A third channel is through a university's medical school and its associated hospital. Because hospitals are open 24/7 and provide one of the most labor-intensive and skill-intensive products that exist, they generate hundreds or even thousands of high-paid local jobs. Much of health care is a local service that follows rather than causes local prosperity. But sometimes hospitals become regional or national providers. Rochester, Minnesota, where the Mayo Clinic is located, Pittsburgh, and Houston attract patients from all over the country and the world. These hospitals are effectively producing a tradable service that is exported outside the local economy—not unlike Microsoft and Apple—and therefore their presence is an important driver of local wealth.

Overall, my research suggests that the presence of a university is on average associated with a better-educated labor force and higher local wages. But at the same time, mayors and local policy-makers should realize that a university—even a good one—is no guarantee of economic success. While most large cities have universities, only a small minority of metropolitan areas have large concentrations of innovative industries. Washington University in St. Louis is a better academic institution than the University of Washington in Seattle, but St. Louis has few high-tech jobs to show for it. In fact, it has been losing population for the past fifty years, while Seattle is now one of the world's most dynamic innovation hubs. Arizona State and the University of Florida are among the largest institutions of higher education in America, but Phoenix and Gainesville rank low on the list of innovation hubs. Cornell and Yale dominate global academic rankings, but other than employers directly connected to these universities, there is little in Ithaca and New Haven to suggest a world-class high-tech cluster.

Thus, proximity to a research university is important, but it is not enough on its own to form a sustainable cluster of innovative companies. This is a key distinction, one that is ignored by countless local governments—from Las Vegas to Detroit, from Italy to China—that invest scarce resources in the creation of research centers. Universities are most effective at shaping a local economy when they are part of a larger ecosystem of innovative activity, one that includes a thick market for specialized labor and specialized intermediate services. Once a cluster is established, colleges and universities play an important role in fostering its growth, often becoming a key part of the ecosystem that supports it and makes it successful.

The Economics of Poverty Traps and Big Pushes

We have just seen the promise and the pitfalls of policies to foster local economic growth by increasing the supply of skilled workers in a city. The alternative approach is to try to increase the demand for labor by attracting employers. This often amounts to offering targeted incentives to innovative companies to locate in a struggling community, in the hope of forming a cluster that in the long run becomes self-sustaining. Once a successful industrial cluster is under way, it tends to strengthen over time. Its labor market and the market for specialized business services become even thicker, and its knowledge spillovers become even stronger. The difficult part, of course, is jump-starting the cluster.

In 2005, Ping Wang was facing a similar problem. He had been chairman of the economics department at Washington University in St. Louis for only a few months, but he quickly realized that the department was in trouble. Wash U. offers an excellent undergraduate program, ranked just below the Ivy League schools, and attracts smart students from all over the country. Unlike its Ivy League competitors, however, Wash U. has never been particularly strong in economics. When Wang became chairman, the department had been stuck for years near the lower part of academic rankings. It simply did not have many professors actively engaged in research.

For a university, there is much to be gained from a strong economics department. Economics tends to be a popular major, because it offers good career prospects. Unlike many other academic fields, economics matters to people outside academia, and economists' research is often cited in newspapers and television. While academic economists' salaries are considerably higher than, say, those of physicists or biologists, economists are cheap for universities, since they do not need expensive labs or sophisticated scientific instruments.

The problem with transforming a sleepy, third-rate economics department into a top-notch research powerhouse, Wang quickly realized, was that academics are kind of like high-tech companies: they tend to be productive and innovative when they are surrounded by smart colleagues with whom they can exchange ideas. Left alone, they tend to stagnate. Thus departments that are already strong tend to become stronger over time, because the presence of productive researchers is attractive to other productive researchers, and departments that are weak tend to get weaker for the same reason. No good academic wants to be the first to move to a weak department. It is not just about prestige; it is about actual productivity.

Economists who study developing countries call this situation a *poverty trap*. Wang realized that there was only one way to break out of this trap: he needed a "big push" strategy. What he did was unprecedented. He called up two academic stars at other universities and made them an offer they simply could not refuse: a yearly salary in the vicinity of $600,000. The gamble paid off. The two stars quit their jobs at more prestigious universities and moved to St. Louis. Now that the department had two stars, other economists began to find the place attractive and accepted Wang's offers there, albeit at normal salaries. As more good economists went there, the department became even more attractive. Wash U. jumped in the rankings. Wang had broken out of the poverty trap. Unfortunately, this positive feedback loop broke down in 2008, when the financial crisis severely diminished Wash U.'s endowment and forced the department to scale back its hiring.

In essence, a city stuck in a poverty trap faces the same challenge. It is trapped by its past. The only way to move a city from a bad equilibrium to a good one is with a big push: a coordinated policy that breaks the impasse and simultaneously brings skilled workers, employers, and specialized business services to a new location. Only the government can initiate these big-push policies, because only the government has the ability to coordinate the individual actors—the workers and employers—to get the agglomeration process going. The idea is to provide public subsidies for those who are willing to move first but then stop the subsidies after the process becomes self-sustaining, in much the same way that Wash U. was willing to overpay the first two people hired but not later ones. The benefits of big-push policies are potentially huge, because declining communities can, in principle, be brought back to life.

But the track record of these policies is mixed. To succeed, the push needs to be *really* big. It also needs to be decisive and sustained, and, most important, the subsidies must target the right beneficiaries. As the story of Wash U. shows, the cost can be high and the success only temporary.

The first and most important big push ever attempted in the history of the United States was the Tennessee Valley Authority (TVA), created in the midst of the Great Depression to lift a desperately poor region out of poverty. According to Franklin Roosevelt—never prone to understatement—the program was intended to modernize the regional economy by "touching and giving life to all forms of human concerns." In practice, this meant investing in large-scale infrastructure programs, particularly electricity-generating dams, whose power was used to electrify the region and boost local productivity; an extensive network of new roads; a 650-mile navigation canal; schools; and flood control systems. A smaller portion of the funds was devoted to malaria prevention, reforestation, educational programs, and health clinics. The scale of the program was enormous, far beyond anything attempted before or since. Between 1933 and 1958, $30 billion from U.S. taxpayers poured into the region. At the program's peak, in 1950, the annual federal subsidy to the region was $625 per household. After 1958 the federal government began to scale back its investment, and the TVA became a largely self-sustaining entity.

This approach to economic development is based on the intuitive notion that public monies can jump-start a local economy trapped in poverty. But critics on both the right and the left have lambasted such initiatives, either as big government overreach or top-down control of local communities. In an influential 1984 article in the *New York Review of Books*, the progressive urban thinker Jane Jacobs wrote a scathing critique of big-push policies, including the TVA, arguing that it is an unnatural way to foster local economies and concluding that "in practice, they work miserably."

How can we assess these place-based policies in a rigorous way? The real test is not whether they create jobs during the push. The fact that an inflow of money temporarily increases economic activ-

ity in an area is hardly a sign that the money was well spent. Instead
we need to look at whether the publicly financed seed can eventually generate a privately supported cluster that is large enough to
become self-sustaining. The idea is that the government-provided
investment carries the local economy past the tipping point but
not any further. At that point the forces of agglomeration take over,
continuing to attract businesses and workers well after the subsidies end.

My colleague Pat Kline and I conducted a careful study of the
TVA initiative and found that the program was successful in generating an industrial revolution in an area that had been largely
rural up to that point. During the big-push years of 1933 to 1958,
manufacturing jobs in the region grew much faster than they did
in the rest of the country, as companies found the cheap electricity
and easy transportation attractive. Manufacturing jobs kept growing faster after the federal subsidies dried up. Even in 2000, more
than forty years after the end of the federal transfers, manufacturing jobs in the region were growing faster than those in comparable parts of the South, although the effect is now slowing down and
will probably disappear soon. While the program was successful in
moving the region from a low-productivity sector (agriculture) to
a high-productivity sector (manufacturing), it did not succeed in
raising local wages in any significant way. The reason is simple: as
more and more jobs were created, more and more workers moved
there from the rest of the South to take advantage of improved
economic conditions. This increase in the supply of labor effectively offset the increase in demand.

The fundamental challenge with policies such as the TVA is
that for them to be successful, local policymakers must be able
to pick promising companies to invest in. They need to be a little like venture capitalists. In this sense, FDR had it easy. When

manufacturing was the engine of job growth and prosperity depended on infrastructure and cheap energy, the recipe for development was obvious. The level of industrial development in the Tennessee Valley was so low that it did not matter too much whether an aluminum smelter, a steel factory, or a chemical factory opened its doors. But today the most important determinant of success for local communities is human capital, and making the right call is much harder. Should a county spend all its money attracting a new nanotech lab, or should it go for Amazon's latest computer farm? A solar-panel R&D facility or a biotech lab? Even professional venture capitalists have a hard time predicting which industries and companies will succeed. For mayors of struggling municipalities, this challenge can prove insurmountable.

Indeed, looking at the map of America's major innovation clusters, it is hard to find an example of one that was spawned by a big push. No local politician set out to create Silicon Valley. And we have seen in the cases of Seattle, Boston, San Diego, and Los Angeles, the success of an original anchor company was typically the seed that grew into a high-tech cluster.

The same is true for smaller, more specialized clusters, arguably a more realistic goal for struggling communities. Consider Portland, Oregon; Boise; and Kansas City, three small high-tech hubs anchored by semiconductors, general high tech, and animal health and nutrition science, respectively. Although small, these are dynamic centers: Portland and Boise produce almost as many patents per capita as Boston. None of these hubs were planned. The opening of Intel's semiconductor facility in 1976 jump-started Portland's high-tech sector. The seed for Boise was planted in 1973, when Hewlett-Packard moved its printer division there. Life science R&D in Kansas City can be traced back to the 1950s, when Ewing Marion Kauffman started his pharmaceutical lab. As pointed out in a recent Brookings Institution study, little of the high-tech presence in these cities resulted from aggressive recruitment of companies by local governments.

Other parts of the world have seen some success. Ireland used a deliberate big-push policy to build up human-capital-intensive sectors that previously did not exist. Through aggressive tax incentives and other enticements, it created important clusters in high tech and finance, although the country's recent financial crisis throws into question the sustainability of such policies. Israel's high-tech cluster, one of the most dynamic in the world, is highly dependent on the country's military. Although the Israeli government did not set out to create a local high-tech sector, its need for innovative defense technologies and specialized human capital indirectly fostered a private sector that later became globally competitive.

Perhaps the clearest example of big-push success is Taiwan, which transformed its rural economy into an advanced one with a dynamic innovation sector through a large-scale policy of government-sponsored research in the 1960s and 1970s. The program succeeded in bringing top Chinese scientists back from the United States and establishing a cluster of publicly supported R&D that eventually became thick enough to sustain private companies. This is one of those rare instances in which policymakers turned out to be good venture capitalists. While they did bet on several failed technologies, they also bet on semiconductors very early on. Semiconductors quickly became the core of Taiwan's high-tech sector and arguably one of its engines of prosperity. More recently, Taiwan's high-tech cluster has been embracing newer technologies, including life sciences. But Taiwan might just be the exception that proves the rule.

Industrial Policies, Green Jobs, and the Challenge of Picking Winners

The story of Fremont, California, illustrates the promise and the pitfalls of big-push industrial policies. As traditional manufacturing continues to shed jobs, some communities have sought to create innovation-driven green jobs in solar and wind power, electric cars, and more efficient batteries. The passing of the baton is nowhere more evident than in the city of Fremont, a mixed-income community that is trying to reinvent itself as a center for green R&D.

Until recently Fremont's economic engine was automobiles. The largest employer in the city was New United Motor Manufacturing Inc. (NUMMI), a large car factory that produced two of Toyota's leading models, the Corolla and the Tacoma pickup truck. Driving along Fremont Boulevard, you pass large factories, warehouses, rail yards full of freight cars, and diesel gas stations for trucks. A local greasy spoon with unwashed windows and a faded yellow sign offers a $3 bacon-and-egg breakfast. Fremont has the classic industrial landscape found in countless communities across America. If it were not for the occasional palm tree, you might think you were in Detroit.

Over the past two decades, with the loss of manufacturing jobs, Fremont's economic future started looking increasingly shaky. When NUMMI finally closed its factory and five thousand jobs disappeared, most residents braced for the worst. But unlike Detroit, Fremont has been able to attract a number of clean-tech companies, some of which have opened in old manufacturing plants. In fact, the old NUMMI factory is now occupied by Tesla Motors, best known for its Roadster, the first serially produced highway-capable electric car available in the United States. For a while the future did not look grim. "You can't throw a Frisbee in Fremont without hitting another solar company," Dan Shugar, the chief executive of the solar power company Solaria, told the press in 2010. At the time, Lori Taylor, Fremont's economic development director, was optimistic that clean-tech companies would eventually become the city's new growth engine.

But in 2011 the city experienced a serious setback when one of its largest employers, a solar panel company called Solyndra, filed for bankruptcy. Solyndra was supposed to be a poster child for enlightened industrial policy, but it has instead turned into a painful cautionary tale. In 2009 the U.S. Department of Energy provided the firm with a loan guarantee of about $535 million, which enabled Solyndra to open a major production facility in Fremont and hire more than one thousand workers to make solar panels. High-profile ribbon-cutting ceremonies followed, together with a visit by President Obama, who confidently presented Solyndra as "the face of a brighter and more prosperous future." Countless news articles touted the promising future of America's "advanced manufacturing." But Solyndra's business model was based on a seriously flawed premise. It depended entirely on the competitiveness of a new type of solar array, which was supposed to generate power more cheaply than silicon-based solar cells. When silicon was expensive, Solyndra's technology seemed inspired. But even then a smart analyst would have realized that the price of silicon was unlikely to remain high forever, because high prices inevitably create a strong incentive for other producers to enter the market, expand supply, and eventually drive down the price. By 2009, when the federal government approved the loan guarantee, the price of silicon was falling precipitously. (Another setback was the decision of the Chinese government to flood the market with heavily subsidized silicon-based solar panels.) In the end, the drop in price caused Solyndra to go bankrupt.

The media frenzy in the wake of Solyndra's collapse focused

mainly on whether political contributions were behind the loan guarantee and overlooked the two most important lessons. First, while one mistake does not condemn an entire program, the track record on industrial public subsidies in the United States and Europe is not great. It is simply too difficult for policymakers, even the brightest and best-intentioned ones, to identify winning industries before they become winners. At the beginning of the last decade, clean tech seemed like an industry on the verge of an explosion. While there has been growth, the employment gains have not been stellar: since 2003, jobs in this sector have grown less than jobs in the rest of the economy. Even if it were clear which industries would drive future growth, it would still be difficult to pick winning companies within those industries. In the Solyndra case, the Department of Energy deemed Solyndra's business prospects worthy of public investment even when its business prospects were rapidly worsening.

The second and most important lesson is that anytime we spend public money to support private enterprises, we need to ask whether there is a sound economic rationale for that decision. Providing early financial support for American solar companies was seen as a way to seed a cluster of companies in the United States that would ultimately attract a growing share of global employment in the industry. In essence, the idea was an industry-wide big push in renewable energy. If an industry is characterized by strong forces of agglomeration and requires a large initial investment, international competition becomes a winner-takes-all race, where the first mover captures all the market. In this case, government subsidies early on could have helped anchor the industry in America. Once national companies gained market share, the subsidies could be removed. As noted by my colleague Severin Borenstein, however, both Germany and Spain had already attempted their own big push in this area. They made enormous investments in the production and installation of solar panels but saw their share decline as soon as the subsidies evaporated. This suggests that the production and installation of panels are not characterized by strong forces of agglomeration—if they were, the industry would be concentrated in Germany and Spain—and therefore a big push does not make much economic sense in this case.

A large array of solar panels covers the entire south side of the roof of my house. In the middle of the day, when the sun is strong, I can see the meter rotating counterclockwise as my panels sell electricity back to the grid. It feels great to contribute to reducing the need for fossil fuels, especially in the middle of the day, when power consumption in California peaks. My panels are ten years old and were made in America. But newer solar panels are increasingly made elsewhere. Many American companies keep their headquarters and research labs in the States but manufacture the panels in such places as the Philippines and China. If this sounds like the story of the iPhone, well, it is.

From the environmental point of view, globalization of the solar-panel industry is good news, as it makes solar energy more competitive with fossil fuels. Buying my solar panels today would cost half as much as it did ten years ago, in part because they would come from Vietnam. From the employment point of view, the real question is what kind of big-push policies makes sense for the United States today. When the government engages in industrial policies, it should mainly be to correct some form of market failure. In the case of green tech, this means externalities associated with the innovation phase, not the production phase. As we will soon discover, there are solid economic (and environmental) reasons for the government to subsidize basic research and applied R&D of green technologies, since this research generates large spillovers. But in light of the experience of Germany and Spain, there seems to be little justification for subsidizing the actual physical production of solar panels.

Where does all this leave Fremont? The city is still courting clean-tech employers and has exempted clean-tech companies from payroll taxes. Although the closing of Solyndra was a major setback, local growth is still driven by R&D-intensive clean-tech companies, along with new biotech and high-tech firms that have also cropped up in recent years. There are now twenty-five clean-tech firms in Fremont, up from six in 2006. While some of them depend at least in part on federal subsidies, others are operating purely on private venture capital. It is still too early to tell whether Fremont's strategy will turn out to be sustainable in the long run. But for now, at least, jobs are being created and the city's prospects look brighter than Flint's or Detroit's.

Bribing Business to Hire Workers

A big-push program the size of the Tennessee Valley Authority would be unthinkable today. But smaller big pushes are very common, in the form of a myriad of federal and state subsidies to attract private investment to struggling communities. Virtually every time a company announces plans for a new headquarters, a lab, or a large production facility somewhere in the United States, the bidding begins. States compete aggressively by offering larger and larger enticements in the form of tax breaks, subsidized loans, local infrastructure, export assistance and financing, workforce training, and area marketing.

These subsidies can be incredibly large. Panasonic recently received more than $100 million ($125,000 per job) to move its North American headquarters to Newark, while Electrolux was given $180 million ($150,000 per job) in tax abatements for its new

establishment in Memphis. Mercedes received a $250 million in-
centive package ($165,000 per job) for locating in Vance, Alabama.
Total state spending on local economic development amounts to
$40 billion a year, significantly more than the cumulative federal
spending for the Tennessee Valley Authority over its thirty years
of government subsidies. These subsidies are one of the rare issues
that Democrats and Republicans tend to agree on. Although their
rhetoric about government intervention sounds different, blue and
red states both engage in efforts to bribe businesses to come to
their jurisdictions. Rick Perry's $200 million "Emerging Technol-
ogy Fund" for Texas companies is similar to initiatives in California, New York, and Massachusetts. The federal loan guarantee program that Solyndra took advantage of was started in 2005 under
George W. Bush and was expanded in 2009 by Barack Obama.

While politicians and the companies they subsidize usually
extol the benefits of these deals, critics complain that they are a
huge waste of public money. Is spending $150,000 per job really
the best way to help the residents of Memphis? What if we just
wrote checks to those residents instead? With two colleagues, I
have studied what happens to local communities when their bid to
attract a large employer by offering subsidies is successful. When
firms are considering where to open a large plant, they typically
begin by looking at dozens of possible locations. They narrow the
list to roughly ten sites, from which two or three finalists are selected. In our study we compared the experience of the counties
that the company ultimately chose (the winner) with the runner-up
counties (the losers). For example, when BMW decided to open a
new plant in the United States in the 1990s, the decision of where
to locate it came down to two finalists: Greenville-Spartanburg,
South Carolina, and Omaha, Nebraska. BMW chose Greenville-
Spartanburg, partly because of an incentives package worth $115
million. In this case and others, the losers were counties that had

survived a long selection process but narrowly lost the competition. They can therefore tell us how the winner county would have fared if it had decided not to bid.

Our data show that in the years leading up to such a bidding war, winners and losers were similar in terms of employment, salaries, and productivity. But afterward, winners experienced a surge in productivity. These productivity gains, which appeared to reflect knowledge spillovers, were particularly large for existing plants that shared similar labor and technology pools with the new plant. We concluded that by making existing producers more productive, a new plant generated an important benefit—a positive externality—for the rest of the establishments in the county. This higher productivity led to more jobs and higher wages. Thus the provision of subsidies might be seen as a way to internalize this externality.

While the theory is clear, in practice these policies do not always work the way they are supposed to. The provision of the subsidy should be commensurate with the magnitude of the social benefit. But when dozens of similar counties are desperate to attract investment from outside, their bids often become so generous that they can exceed the social benefits to the community. Mayors and governors have an incentive to bring the new company to town, no matter what the cost. When they are successful, front-page stories in local newspapers tend to focus on the hundreds of future local jobs, not on the fine print of the financial packages offered. When they are unsuccessful, local politicians are lambasted for not doing enough for the local economy. All of this can lead local governments to overbid. In such a case, the only winners are the owners of the company being courted, since state and local governments end up stuck with the bill. And even when these subsidies make economic sense for a particular county, they do not always make sense for the country as a whole, as competition among municipalities for a given company can turn out to be a zero-sum game for the nation.

Empowering Neighborhoods

One example of a place-based policy that has been successful at helping struggling communities create jobs and raise salaries is the Empowerment Zone Program. Created in 1993, during the first Clinton administration, the program provided a package of employment tax subsidies and redevelopment funds to "distressed" urban areas. Instead of targeting entire regions, the program zeroed in on impoverished neighborhoods in Atlanta, Baltimore, Chicago, Detroit, New York City, Philadelphia, Los Angeles, and Cleveland. The federal subsidies were designed to encourage economic and social investment. They paid for jobs and training programs for individuals who lived and worked in these poor communities as well as business assistance, infrastructure investment, and neighborhood development. One of the most visible examples of a neighborhood that received the Empowerment Zone funds is the area around 125th Avenue in Harlem, which went from being a problematic and crime-ridden neighborhood to being one of the liveliest parts of New York City.

In 2010 a team of three economists led by Pat Kline completed an in-depth evaluation of the program. They compared areas that were designed as Empowerment Zones to equally poor areas that were nominated by their local governments but never received funding. Their findings were encouraging. In the first five years of the program, the Empowerment Zone neighborhoods experienced substantial increases in local jobs—about a 15 percent gain over the other neighborhoods. The hourly wages paid to residents also appeared to increase significantly, by approximately 8 percent.

Why was this program successful when many other place-based policies have failed? The process of neighborhood revitalization, just like the growth of an industrial cluster, generates many externalities. For example, a boarded-up property that turns into an active retail store brings more business to nearby stores, increases foot traffic, and reduces loitering and crime in the entire block. A business that improves its façade benefits not just its own building but all the properties around it. A new job in a tough neighborhood means more than just a job for one worker; it creates social benefits in the form of reduced public assistance and reduced crime. Dealing with these externalities is the secret of successful local economic development. Certain actions will benefit the entire community, but individuals will not take them on their own, because although social gains are large, private gains are limited.

In the case of Empowerment Zones, the government was able to incentivize these actions. Although politically the program was sold only as a way of transferring financial resources to needy urban residents, the Empowerment Zones program was successful because it solved this collective action problem. A second reason was that, unlike industrial policies that target specific companies or sectors, policymakers did not act as venture capitalists: public funds were directed toward any form of investment that might benefit the community. Third, and crucially, public subsidies were not just a giveaway but were the catalysts for significant *private* investment. By one estimate, for each dollar in public funds that was spent, three and a half dollars were invested by private businesses, thus creating a virtuous cycle of growth. This is the true hallmark of a successful place-based policy, because it indicates that when the public money ends, the area will keep creating jobs. Finally, incentives were targeted and well designed. In contrast to the typical state subsidy to new manufacturing establishments, which end up shifting investment from one county to another, the subsidies in this program targeted residents of areas with high unemployment, and thus most of the jobs created did not come at the expense of job creation in other areas.

What about gentrification? The program did not seem to lead to a displacement of original residents, because rents in the area were not significantly affected. (Harlem is a major exception.) This probably reflects the fact that the program targeted workers who already lived and worked in the neighborhood.

Overall, the program was a good investment for the government. It generated annual wage increases worth a total of approximately $900 million per year. Assuming that most of the jobs created went to previously unemployed workers, the annual return on the taxpayer money invested was 15 percent. At the same time, the program helped residents of some of the poorest urban communities in the country by adding jobs, raising local incomes, and internalizing the inevitable externalities associated with revitalization.

Change is always painful. Economic change is particularly painful. People invest time and energy in jobs, careers, and communities. Asking them to change because the economy around them has changed is asking a lot, but in some cases there simply aren't viable alternatives.

People often have unrealistic expectations of their governments. The role that local governments can play in revitalizing struggling communities is less extensive than most voters realize and most mayors would like to admit. The reality is that a city's economic fate is in no small part determined by historical factors. Path dependency and strong forces of agglomeration present serious challenges for communities without a well-educated labor force and an established innovation sector. Local governments can certainly lay a foundation for economic development and create all the necessary conditions for a city's rebirth, including a business climate friendly to job creation, but there is no magic formula for redevelopment. Like politics, all innovation is local: each community has its own comparative advantage. Local governments must build on their existing capabilities by leveraging local strengths and expertise. The use of public funds to create jobs must be reserved for cases where there are important market failures and a community has a credible chance of building a self-sustaining cluster. Ultimately, though, local policymakers should realize that when it comes to local development, there is no free lunch.

7

◆

THE NEW "HUMAN CAPITAL CENTURY"

IN THE PAST, good jobs and high incomes were tied to the large-scale production of manufactured goods. Factories were the places where economic value was created. But today little value remains in the production of goods that anybody can make. Good jobs and salaries increasingly come from the production of new ideas, new knowledge, and new technologies. This shift will continue and probably accelerate in the future. In the coming decades, global competition will be about attracting innovative human capital and innovative companies. The importance of geography and the forces of agglomeration in determining the location of human capital will keep growing. The number and strength of a country's brain hubs will determine whether it will prosper or decline. Physical factories will keep losing importance, but cities with a large percentage of interconnected, highly educated workers will become the new factories where ideas and knowledge are forged.

Is America ready? Yes and no. The country's economy is strong in many ways. Its labor market remains among the most efficient, flexible, and meritocratic in the world. It rewards personal effort and risk-taking more effectively than most other countries. This is a crucial advantage in attracting top talent. Moreover, for all our recent problems, America's capital markets and venture capital system remain among the most efficient anywhere, providing financing to innovative entrepreneurs with good ideas and a willingness to work hard. As we have seen, this makes a huge difference for the creation of local businesses. Most important, because of its dynamic brain hubs, the United States is well positioned to continue to generate innovative activity.

At the same time, our economy has some serious flaws. These flaws are probably not too surprising, because they are to some extent a reflection of our national character. As a society, we are much too focused on the present at the expense of the future. Our culture glorifies instant gratification and quick results, and it shuns long-term commitment. Most of the energy and attention in our policymaking is concerned with short-run issues, such as how to stimulate the economy over the next six months or how to deal with this week's employment numbers. While short-term issues can be pressing, their importance pales relative to that of long-term ones, because the latter are the ones that really affect our standard of living in profound and permanent ways. The magic of compound growth means that even tiny differences in growth rates can have enormous consequences for our future jobs and incomes. Thus, policies that can increase growth even marginally are vastly more important than any short-term fix to the economy. Our ethos of immediate reward and our almost structural inability to take responsibility for long-term problems is leading us to underinvest in our future. If left unchecked, this tendency could have truly disastrous consequences. The effects will be felt most strongly by our children and by the least fortunate in our society.

In particular, two structural weaknesses that have emerged in the past three decades severely limit America's economic potential and create serious social imbalances in our country. Human capital and research are the engines that sustain the American economy and its workforce. As we have seen, some American cities are lacking in both. But the problem is bigger than that. The United States as a whole is not investing enough in human capital and research. As a consequence, our salaries are not growing at the rate they used to, and inequality is increasing.

In Chapter 6 we examined the role of place-based policies in helping to level the playing field. In this final chapter we look at what America as a nation should be doing to regain the strength that made it the dominant economic player of the twentieth century. These reforms are crucial for our future as a society, because they will affect our capacity to grow for decades to come.

The Social Return of Research

The first problem with the American economy is that public and private investments in research are insufficient. This is not to say that American universities and companies invest too little relative to other countries (although this is sometimes true) but rather that they invest too little compared to what would be socially optimal. This is caused by a serious failure in the market for knowledge. As we saw in Chapter 4, the existence of significant knowledge spillovers means that the creators of new ideas are not always fully compensated for their efforts, as some of the benefit of their research inevitably accrues to others. This is not just an American problem, but it is more salient for the United States than for other countries because of the role that innovation will play in our future growth.

Academia has traditionally provided the basic science upon

which the private sector builds new commercial applications. This is a big reason that the federal government subsidizes academic research through institutions like the National Science Foundation and the National Institutes of Health. The problem is that this funding has not kept up with the increased value of knowledge. Globalization and technological change have resulted in increased returns on the creation of new commercial knowledge, as we have seen. This means that the potential economic value of new discoveries in basic science has also increased. If the return on an investment increases, the rational reaction is to invest more. And yet the resources that the federal government devotes to supporting basic research have actually declined. As we saw in the case of solar energy, federal and state governments generously support factories and production facilities with dubious potential but devote far fewer resources to foundational science.

Knowledge spillovers do not flow only from academia to private companies. The most important kind flows between private companies. Innovative companies that invest in research appropriate just some of the benefits of their efforts. Take a firm that develops a new energy-efficient battery for electric cars. A patent will give this firm a claim to the profits from its new technology. But when the patent is filed, all other firms in the industry can see the breakthrough that led to this innovation, and this knowledge is likely to inspire new ideas for related technologies or products.

The development of any new product generates similar spillovers. Consider, for example, the introduction of the iPad. Because the product was completely new, nobody really knew its market potential. Apple carried substantial risks, because it had invested significant resources in the iPad's development. Indeed, when Steve Jobs unveiled the device in front of a select group of journalists and opinion leaders, including Al Gore, at an invitation-only event in San Francisco in January 2010, many industry analysts were skeptical, arguing that the iPad was just an expensive gadget and therefore destined to remain a niche product. Some ridiculed it as an outsized iPhone without the phone and predicted that it would generate little interest. After the launch, however, it became clear that the iPad was going to be an international sensation, and many competitors immediately started developing their own versions. Essentially, those competitors benefited from the information generated by Apple's risk-taking.

In practice, the magnitude of these sorts of knowledge spillovers is substantial. In one of the most rigorous studies to date, two economists—Nick Bloom of Stanford and John Van Reenen of the London School of Economics—followed thousands of firms between 1981 and 2001 and found that the spillovers were so large that R&D investment of one firm raised not only the stock price of that firm but also the stock price of other firms in the same industry. Part of the spillover is global in scope. For example, an increase in R&D investment by U.S. firms in the 1990s translated into significant productivity increases for UK firms in similar industries, with the majority of the spillover accruing to firms with an American presence. But a significant part of the spillover is local, because it occurs between firms that are geographically close. So new knowledge generated by American companies benefits other American companies.

In essence, private investment in innovation has a private return for the firm that makes that investment, but it also supplies a social return that benefits other firms. The problem is that the market provides less investment in innovation than is socially desirable, because the return on such investments cannot be fully captured by those who pay for it. The only way to correct for this market failure is for the government to step in and compensate those who invest in R&D for the external benefits that they generate. This is the main reason that the United States government, as well as governments in most industrialized countries, subsidize R&D through tax breaks. It is important to realize that this is not about fairness — it is purely about economic efficiency. The government does not subsidize innovators because it has a moral obligation to do so. It subsidizes innovators because it is in the interest of the American economy to do so.

The problem is that the difference between private and social return on innovation is much larger than the current subsidies. Bloom and Van Reenen estimate that the social rate of return on R&D is about 38 percent, almost twice as large as the private return. The implication is jarring. The United States is not just underinvesting in R&D; our current level of R&D investment is barely half of the socially optimal level. The lessons for economic policy are clear: the current U.S. tax credit for corporate spending on R&D is far smaller than it should be. We need to increase federal support for academic research in science and engineering and especially for private R&D. This is a solid investment that will ultimately pay for itself. Other levels of government should do their part too. Because the benefit of spillovers is in part local — helping some communities but not others — the efficient distribution of cost is one in which state and local governments also contribute to the subsidy.

Interestingly, not all innovators deserve the same level of subsidization. When Bloom and Van Reenen zoomed in on specific companies, they found that some generate more social returns than others. Computers and telecommunications companies generate larger social returns than the social returns for pharmaceutical companies. R&D in the pharmaceutical sector often replicates what other firms are already doing, and it is not uncommon for several companies to be racing to patent a drug to treat the same condition. In this competition, the winner reaps the benefits and all the others end up having wasted precious resources. This business-stealing effect tends to reduce the social value of pharmaceutical R&D and suggests that R&D tax credits should be smaller in this industry.

The second fundamental problem with the American economy is significantly more difficult to address. America does not create enough human capital. Over the past thirty years the United States has failed to raise its percentage of college-educated young adults substantially. Companies—especially those in innovative industries—are finding it increasingly hard to hire employees with the right skills. And workers are experiencing a steep increase in income inequality. Both problems reflect a serious imbalance in America between the demand for human capital and its supply.

We have heard a lot of talk recently about America's education crisis. But the argument I am making here is not only a moral one, although I do believe that we should seek to give all our children access to a first-rate education. It's a pragmatic one. It's about whether we want the third America—the America of hollowed-out urban cores, high crime, low wages, and short life spans—to be the *only* America.

Why Inequality Is About Education

For Europeans who visit this country, one of the most refreshing and inspiring aspects is the absence of a strong sense of social class. Even today the idea of class is pervasive in countries such as Great Britain and (as much as its inhabitants would bitterly deny it) France. Blue-collar workers have a very different perception of their place in society from white-collar professionals. This of course affects not just their sense of self but also their aspirations and political inclinations. By contrast, the concept of class does not resonate here. In fact, when asked by pollsters, most Americans—those making $20,000 and $300,000 alike—answer that

TABLE 4: HOURLY AVERAGE WAGE OF MEN, BY EDUCATION (2011 DOLLARS)

	1980	2010	Percent change
Dropout	$13.7	$11.8	-14%
High school	$16.0	$14.8	-8%
College	$21.0	$25.3	+20%
Advanced degree	$24.9	$33.1	+32%

Note: Data include all full-time workers aged 25–60.

they belong to the middle class. I have always thought that this is one of the fundamental cultural differences between the Old and New Worlds, a difference that could account for the stronger entrepreneurial spirit among Americans and the different attitudes toward income inequality and income redistribution.

But whatever Americans' self-perception is, differences in income levels are growing. As we have seen throughout this book, this increase has a strong geographical component. But it is also skill-based. Table 4 shows how the hourly wage of full-time male workers has changed since 1980 depending on their level of schooling. The wages of men with less than a high school education and of those with just a high school education today are lower than they were in 1980. By contrast, the wages of college graduates have increased significantly. The gain is even larger for workers with a master's degree or a PhD.

The "college premium"—the wage gap between those with high school and college educations—is the measure that labor economists most commonly use to track changes in labor market inequality, because it best captures the difference between the typical skilled worker and the typical unskilled worker. This premium was relatively small in 1980—only 31 percent—but has been growing every year since then and is now more than double its 1980 level. This difference is even higher when you account for other aspects of compensation, as college graduates tend to have better employer-paid health insurance and more generous pension contributions.*

Wage inequality is a hot topic right now. One widespread misconception is that the problem of inequality in the United States is all about the gap between the top one percent and the remaining 99 percent. Although the super-rich capture people's imaginations, their earnings are not the main driver of these figures. These numbers do not reflect the gap between the millionaires with penthouses on Park Avenue or the startup whiz with millions in stock options and the rest of us. Instead they reflect the difference between the typical college graduate and the typical high school graduate: regular people with regular jobs, families, and mortgages. If we dropped all the CEOs and financiers from the data, the table would be largely unchanged. The most important aspect of inequality in America today is not what happens to a few thousand tycoons. The increase in their share of wealth is certainly a problem, but not as consequential as the rapidly growing divide between the 45 million workers with a college education and the 80 million workers without one. As we are about to discover, this is the difference that really matters for people's lives—their standard of living, their family stability, their health, and even the health of their children.

Another misconception is that the increase in wage inequality is mostly caused by deliberate economic policies: the decline in the real value of the minimum wage; the weakening of the insti-

---

* In addition to a marked increase in inequality across educational groups, there has been a significant rise in inequality *within* educational groups. For example, among college graduates, the distance between those with high wages and those with low wages has increased sharply.

tutions that used to protect low-wage earners, such as unions; and the general trend toward deregulation. But a careful reading of the data suggests that institutional factors have played only a secondary role. Wage inequality has increased in the past thirty years in many different countries in Europe, Asia, and the Americas — each with different labor market institutions, regulations, tax policies, union penetration, and levels of minimum wage. In the United States, wage inequality has increased both in blue states with high minimum wages and in red states with low minimum wages. It has increased in most sectors, both those with high unionization rates and those with low unionization rates.

The reality is that the trends in wage inequality reflect forces that are deeper and more structural. A vast body of recent research indicates that these trends can best be explained by changes in supply and demand—namely, an increase in the demand for college-educated workers and a slowdown in the supply. In a remarkable book, the Harvard economists Larry Katz and Claudia Goldin explore this race between demand and supply over the course of the twentieth century and demonstrate that for most of the century, supply outpaced demand, and this kept inequality in check. The share of Americans going to college was growing at a rapid rate during the 1950s and 1960s, and as a consequence the earning gap between college and high school graduates was stable or declining. But over the past four decades demand has prevailed and inequality has exploded. The slowdown of supply has been particularly pronounced for men: between 1980 and today, college graduation among young white male adults (ages twenty-five to thirty-four) rose very little, from 22 percent to 26 percent. Fortunately, the numbers for women look better: although they too initially slowed, they have recently picked up again. Today 60 percent of recent college graduates are female and 40 percent are male, a remarkable change since 1980, when the reverse was true. The two

Harvard economists show that if the increase in the number of college graduates since 1980 had kept pace with the earlier rate, wage inequality in America would have fallen in the past thirty years, not increased.

What should we do about this? We know the reasons for the increase in the demand for skilled labor: technological progress, globalization, outsourcing, and the shift away from traditional manufacturing industries. And we have seen why the transformation to an idea-driven economy fueled by human capital is a good thing for America. Thus there is little the government could or should do to limit the increasing demand for skilled labor. On the other hand, there is quite a bit we can do to increase its supply.

To step back for a minute, the slowdown in the supply of skilled labor is rather baffling. Given that the wages of college graduates have increased so much more than those of less educated workers, why aren't more young people taking advantage of this fact by going to college in the first place? When presented with this puzzle, the standard reaction is to point to the growing cost of college: college was cheap in the 1970s, but since then tuition at both private and public institutions has skyrocketed. Tuition at Yale has increased from $6,210 in 1980 to $40,500 today. Tuition at Berkeley has increased from $776 in 1980 to $13,500 today, an even steeper increase percentage-wise. These are not exceptional cases; tuition at the typical American college has increased tenfold in the past three decades, substantially more than most other goods or services sold in the economy. Is this the root of the problem?

Most seventeen-year-olds think about going to college as a way of moving out of their parents' house, having new experiences, and maybe occasionally getting drunk. Economists, undeniably a dull bunch, tend to think about the decision in purely financial terms. In 1964 the University of Chicago economist Gary Becker wrote a book called *Human Capital*, for which he was later awarded the

Nobel Prize. The central idea is simple but powerful. The decision to go to college is at its core just like any other investment decision. When you buy Treasury bonds, you pay a cost up front and you receive a flow of revenues over time. Becker pointed out that going to college is much the same. In 2011 the up-front cost was remarkably high. Taking into account tuition costs and the forgone salary that an individual would earn by working for four years, the total investment cost comes to $102,000.

This is an enormous sum. But the benefits are much larger. Figure 11 compares the average salary of a typical college-educated worker with the salary of a worker with a high school diploma over their lives. The gap is large at age twenty-two and grows larger over time. It reaches a peak at age fifty, when the average college graduate makes almost $80,000, compared to $30,000 for the average high school graduate. If a seventeen-year-old decides to go to college, she can expect to earn more than a million dollars over her lifetime. If she does not go to college, she will make less than half that amount.*

---

* These differences are enormous, but are they real? Is someone who went to Princeton really comparable with someone who took a job in the local supermarket right after high school? The concern is that the higher salary of college graduates creates a statistical illusion that reflects self-selection, making the real return much lower. College graduates tend to have higher IQs and better-off parents with more family connections than those who do not go to college. All of these factors indicate that even without a college degree, those in this group would have had higher salaries. Determining the importance of selection is crucial: if college graduates have higher salaries not because of their education but because they have higher IQs or more family connections, then it would make no sense to pay all that money for college. Several studies have investigated this question. To control for self-selection, they look at cases in which a new college opens in a town or the state suddenly decides to offer more financial aid. In these cases, the self-selection problem is less severe, because the decision to go to college is driven by external factors. The overwhelming conclusion of these studies is that while it is true that those who go to college tend to have more analytical ability than those who do not, college does directly raise people's productivity and salaries.

Of course this does not mean that everyone should go to college. The history

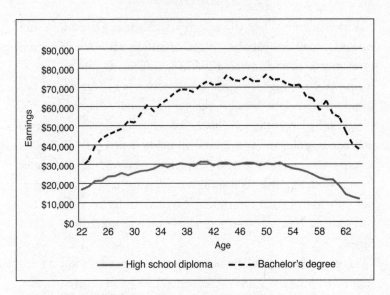

Figure 11. Annual average earnings by age and education
Source: *Adapted from Greenstone and Looney, www.brookings.edu.*

Not only is college a good investment, it is one of the best investments around. Let's say the parents of our seventeen-year-old are not fully convinced. Instead of paying $102,000 for her college education, they give her $102,000 in stocks or bonds. Would she be better off going to college or enjoying the return on her finan-

---

of technology is dotted with brilliant innovators who dropped out. If you have a groundbreaking idea, it obviously makes sense to pursue your dream. When Bill Gates decided that working for Microsoft was more important than finishing Harvard, it was a turning point for the software industry. If he had finished college first, Microsoft might not have become the dominant force it later became, and Gates himself would have been several billions poorer today. If Mark Zuckerberg had stayed in college, his nemeses, the Winklevoss twins, might have developed a social networking site before him. In 2010 the venture capitalist Peter Thiel launched a controversial charity that offers $100,000 scholarships to twenty-year-olds with exceptionally promising business ideas to enable them to drop out of college. But these are all exceptions to the rule. For the average individual, education pays, and today it pays more than ever.

cial investment? When the researchers Michael Greenstone and Adam Looney compared a college education with other financial investments, they discovered that it is difficult to find an investment that has a higher return. Investment in a college degree delivers an inflation-adjusted annual return of more than 15 percent, significantly larger than the historical return on stocks (7 percent) and bonds, gold, and real estate (all below 3 percent). College is where smart investors should put their money. And this is without even considering risk. As it turns out, investment in human capital not only has a higher return but also tends to be safer than other investments. If college were a stock, it would be the darling of Wall Street.

It gets even better. The benefits of a college education are not limited to financial gains—they extend to health, marriage, and many other aspects of life. In work based on a representative sample of more than 2 million mothers, Janet Currie and I found that better-educated mothers are more likely to be married: 97 percent of college-educated mothers were married at the time of delivery, while only 72 percent of high school dropout mothers were married, and among the married, the former group had husbands with significantly higher earning potential than the latter group. A good education benefits not just the receiving individual but also her children. We found that only 2 percent of college-educated mothers smoked during pregnancy, compared with 17 percent of mothers with a high school education and 34 percent of mothers who dropped out of high school. College-educated mothers are significantly less likely to have children who are premature or have low birth weight, two important predictors of later health problems. Children of college-educated parents are not just healthier at birth but also tend to have more and better schooling themselves, further increasing their health and income potential. It is important to note that these differences are not just correlations explained by the fact that women who go to college have developed good habits before going, as a result of upbringing or other socioeconomic factors. Rather, these differences reflect causation, since we observed large gains for women right after a new college opened in their county of residence but not before.

In work with Lance Lochner, I have found that education has an additional benefit: it lowers the probability of being involved in criminal activities. Among white males, the likelihood of being incarcerated for committing a crime drops significantly for individuals with more schooling, and the effect is even larger for African Americans. Interestingly, this is a case where education benefits not just the individual who acquires it but society as a whole.

Education clearly does many wonderful things for individuals, families, and communities, and it is probably one of the best investments out there. So the fact that more American teenagers aren't going to college is truly surprising. What factors are limiting the expansion of human capital in America?

One obstacle is the fact that many families simply do not have and cannot borrow enough money to cover the up-front costs of an education. Typically, when someone has a good investment idea but no cash, he goes to a bank to ask for a loan. Each year millions of small businesses are born this way. But this is where investment in human capital differs from other kinds of investment. Starting a business usually involves investing in goods that can be used as collateral, such as machines or real estate. Human capital, in contrast, is completely immaterial. This explains why the private sector isn't jumping in to help people go to college. Think about a hedge fund that, instead of investing in existing companies, invested in the education of young people in exchange for a portion of their higher salaries when they finally entered the labor market. A social enterprise called Lumni is trying to do just that. It has raised $15 million to finance the education of low-income students in the United

States and Latin America. But without collateral, it is difficult to see how this business model can be scaled up. Lumni does require its students to repay a fixed portion of their income for ten years after graduation, but this is largely a moral obligation, as in practice it would be costly and time-consuming to enforce it. Private-sector involvement remains limited, which is why the government offers subsidized loans to deserving students. Economists disagree on exactly how big a role credit constraints play in the low college enrollment among less wealthy families, but it is clear that for many families they are a significant factor.

A second stumbling block is the country's increased geographical segregation along educational lines. As we have seen, families in which the adults are college graduates increasingly live near other college graduates, and families in which the adults are high school graduates increasingly live near other high school graduates. This matters for the children, because peer effects are probably an important determinant of college enrollment. Children who grow up with friends who have no interest in college are less likely to go than children with college-bound peers. This "social multiplier" inevitably exacerbates the differences in education among different socioeconomic groups.* And this brings us to the most important factor: early education. The Nobel laureate James Heckman has long argued that the accumulation of skills is a dynamic process: "Skill begets skill. Early investment promotes later investment." To really understand the lack of investment in college, we have to go

---

* A third factor is the higher cost of living faced by college graduates, which we considered earlier. We saw that the jobs for college graduates tend to concentrate in expensive metropolitan areas while the jobs for high school graduates tend to concentrate in more affordable areas. This makes college investment somewhat less attractive than it would appear, although still very well above alternative investments.

back to high school, and probably even further. If we don't start investing in our children earlier, college will remain a distant dream.

Math Races

During a recent visit to a Silicon Valley high-tech firm, I asked the Indian-born engineer and company executive Nimish Modi why the majority of engineers in his firm were foreigners. He replied, "The shortage of U.S.-born engineers goes all the way back to high school. It is all about American high schools and the lack of focus on technical education." To his dismay, he added, "I am having the same argument every day with my own son, who was born here and has no interest in math."

To see exactly how big the problem is, consider the results of the Programme for International Student Assessment (PISA), a standardized math and science test administered to fifteen-year-olds in seventy countries. Because all students face the same set of questions, the test is a good way to compare the quality of math and science education around the world. (Math and science are arguably two of the key ingredients in innovation.) The results, reported in Table 5, are rather troubling. At the top of the list is Shanghai (which was tested separately from the rest of China), immediately followed by Finland, Hong Kong, Singapore, and several rich countries, including Japan, Canada, and Australia. This is in itself a stunning realization. Although Shanghai belongs to a developing country, students there perform better than those in all the wealthy nations of the world, including well-organized, homogeneous northern European societies that invest substantially in education. In the group just below the top countries are other European nations, including the Netherlands, Germany, and the United Kingdom. At the bottom of the table are poor countries such as Tu-

TABLE 5. PISA SCORES FOR SELECTED COUNTRIES, 2009

	Math	Science
Shanghai-China	600	575
Finland	541	554
Hong Kong	555	549
Singapore	562	542
Japan	529	539
South Korea	546	538
New Zealand	519	532
Canada	527	529
Estonia	512	528
Australia	514	527
Netherlands	526	522
Liechtenstein	536	520
Germany	513	520
Chinese Taipei	543	520
Switzerland	534	517
United Kingdom	492	514
Slovenia	501	512
Macao-China	525	511
Poland	495	508
Ireland	487	508
Belgium	515	507
Hungary	490	503
United States	487	502
Czech Republic	493	500

	Math	Science
Denmark	503	499
France	497	498
Iceland	507	496
Sweden	494	495
Austria	496	494
Portugal	487	493
Italy	483	489
Spain	483	488
Russia	468	478
Greece	466	470
Turkey	445	454
Chile	421	447
Serbia	442	443
Bulgaria	428	439
Romania	427	428
Uruguay	427	427
Thailand	419	425
Mexico	419	416
Brazil	386	405
Colombia	381	402
Tunisia	371	401
Argentina	388	401
Indonesia	371	383
Peru	365	369

Source: Programme for International Student Assessment

nisia, Peru, and Indonesia. The United States is in the middle of the pack, squeezed between Hungary and the Czech Republic and well below Poland, Slovenia, and Taipei. In math the United States is actually closer to the bottom than to the top. Similarly disappointing results can be seen in literacy and problem-solving tests. It is not just the quality of schooling that is poor. Compared with thirty other developed countries, the United States ranks eleventh in the amount of schooling achieved, according to the OECD.

Equally troubling is the degree of inequality in PISA scores. The United States is one of the nations with the largest gap between students at the top and the bottom, alongside Brazil, Indonesia, Mexico, and others. Not only does the American system fail the average student; it fails disadvantaged students the most. This undermines the social contract on which our society is based: the implicit promise of equal opportunity, irrespective of family background.

One common reaction to these troubling numbers is to pretend that they don't matter very much. After all, the United States still has the best research universities in the world. After all, the innovation sector ultimately depends on the breakthroughs of college graduates and those with master's and doctoral degrees. But this view is shortsighted. In a series of research articles, Heckman has argued that the high school dropout rate in this country accounts for a substantial portion of the recent slowdown in the growth of college-educated workers. Throughout the first half of the twentieth century, each new generation was more likely to graduate from high school than the previous one. The high school graduation rate peaked at around 80 percent in the late 1960s. Incredibly, it then started to decline. Although the decline has been arrested, the rate today is not noticeably higher than it was in 1970. That year, we had the world's highest high school (and college) graduation rates. Today we have been surpassed by scores of other nations. Basically, as far as high school is concerned, the past four decades have been a gigantic lost opportunity. Heckman writes, "The decline in high school graduation since 1970 has flattened college attendance as well as growth in the skill level of the U.S. workforce. To increase the skill levels of its future workforce, America needs to confront a large and growing dropout problem." This increase in dropout rates is likely to result in lower productivity and more inequality in the future. It is clear that unless the United States substantially increases the number of skilled workers with a college education, the supply of human capital in the country will not meet demand, and inequality in our society will keep increasing.

Social imbalance is not the only problem. It is more and more challenging for innovative companies to find workers with the right skill set. Not only are we not producing enough college graduates, we're not producing the right kinds. One economist has pointed out that while the number of students in visual and performing arts has doubled in the past twenty-five years, the number of students in computer science, chemical engineering, and microbiology has remained flat or declined. In a recent meeting with analysts, the CEO of 3M, George Buckley, complained that he needs to hire more scientists and engineers but that it would be impossible for 3M to find them only among Americans. The CEO of Caterpillar, Doug Oberhelman, was even more explicit: "We cannot find qualified hourly production people, and for that matter, many technical, engineering service technicians . . . The education system in the United States basically has failed them, and we have to retrain every person that we hire," adding that this is clearly hurting the productive base of America. In one recent survey, 47 percent of biotech CEOs said that the lack of skilled workers is one of their top three concerns for the future. I have heard the same complaint from countless Silicon Valley human resource managers. Given the PISA scores, it is surprising that Silicon Valley is not situated in Singapore or Slovenia rather than California.

There are two ways to increase human capital in America. One way is to dramatically improve the quality of education—particularly high school math and science—in order to increase the number of Americans with college degrees. The other is to import human capital from abroad by allowing skilled immigrants to move here. Both will do the job, but they involve different financial commitments and have profoundly different implications for the future of American society.

Ethnic Inventors

The percent of smart and highly motivated individuals born in the United States is the same as the percent born elsewhere in the world. But the percent of smart and highly motivated individuals who work in the United States is probably higher. One of America's crucial advantages has always been its ability to attract ingenious and ambitious foreigners to its shores. This ability is largely cultural, as few other societies have been so open to absorbing so many foreigners. But it is also solidly rooted in economics. Although the United States has effectively given up on properly educating its indigenous workforce, it still rewards skills, which is why talented immigrants come to the United States. America's innovation hubs are a magnet for hardworking foreign-born entrepreneurs and scientists, and this reflects the fact that in the United States, outstanding workers—those who put in more effort to get better results, or who are more creative and generate better ideas—are likely to be recognized and rewarded.

This is not always the case in other countries. For example, workers in continental Europe experience very different incentives. Rigid union contracts constrain salaries, work schedules, and promotions. Decision-making is hierarchical, and promotions often depend on seniority. Advancement takes considerably longer (as does demotion). Salary differences are smaller, but effort and ability are less likely to be rewarded. If you are a below-average worker, Europe offers better security. If you are an exceptionally talented individual, however, the United States offers more: your career will progress faster and your salary will be significantly higher. The United States has one of the highest returns on education among industrialized countries. This is true before taxes, and even more so after taxes, because tax rates are higher in most other countries. Clearly the American labor market is very attractive for the best and the brightest.

Therefore, it should not be surprising that those who come to America are increasingly skilled. The percentage of immigrants who are college-educated and who have a postgraduate education (master's degree or PhD) has risen considerably since 1980. This trend accelerated in the past decade, when the cumulative number of low-skilled immigrants dropped and the cumulative number of highly skilled immigrants remained strong. (In the recession years 2008–2011, the size of both groups declined substantially owing to the general weakness of the American labor market.) Today an immigrant is significantly more likely than a native-born American to have a master's or doctoral degree, a fact that is often lost in the immigration debate, and has the same probability as a native-born American of having a four-year college degree. (However, immigrants are also considerably more likely to have very low levels of schooling. Thus the immigrant population is significantly more polarized than the native population.)

While the percentage of skilled immigrants is increasing nationwide, this trend looks vastly different in different parts of the country. Some cities are magnets for highly educated professionals with graduate degrees, while others are magnets for laborers who never went to school. In the first group, immigrants tend to be more educated than natives, so their arrival increases the average human capital in the community. In the second group, immigrants tend to be less educated than natives, so their arrival dilutes the average human capital. In the middle are cities where immigrants and natives have a similar level of education. Just as there are three Americas for natives, there also are three Americas for immigrants.

Map 5, from a Brookings Institution study, identifies cities in which college-educated immigrants outnumber high school dropout immigrants by at least 25 percent and cities in which the reverse is true. New Haven, Minneapolis, Philadelphia, San Francisco, Washington, D.C., and New York are examples of destinations for high-skilled immigrants. Surprisingly, this group also includes some former manufacturing hubs, like Pittsburgh, Albany, Buffalo, and Cleveland. Overall, forty-four metropolitan areas attract mostly high-skilled immigrants, while thirty attract mostly low-skilled immigrants. This latter group includes cities near the border in the West and Southwest, such as Phoenix, Arizona; Bakersfield, California; El Paso, Texas; and McAllen, Texas, where the majority of immigrants have only a few years of schooling or no schooling at all. It also includes some large cities in the Great Plains, such as Oklahoma City, Omaha, Tulsa, and Wichita. These differences between cities have been growing for the past three decades. Remarkably, even in terms of immigration, a great divergence is taking place.

These vast geographical differences are both a cause and an effect of the geographical differences in the innovation sector. Innovation hubs attract highly educated immigrants because they have jobs for them. At the same time, by increasing the local human capital, highly educated immigrants make innovation hubs more productive and creative. In 2010, Jennifer Hunt and Mar-

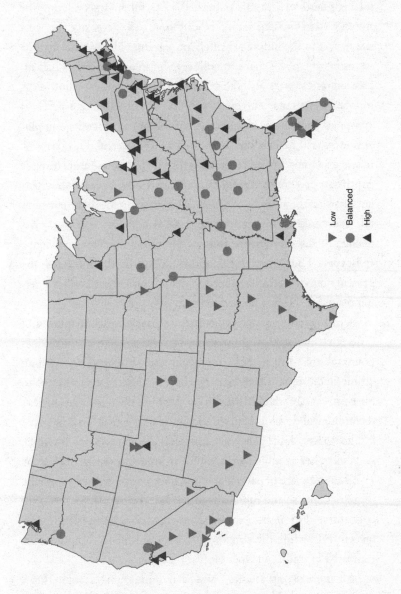

Map 5. *The educational level of immigrants, by metropolitan area*

Source: Hall, Singer, De Jong, and Roempke Graefe, *www.brookings.edu.*

jolaine Gauthier-Loiselle published a careful analysis relating the number of patents generated by different U.S. states over the past sixty years to the inflow of skilled immigrants. They found that the states in which the share of college-educated immigrants had increased fastest were also the states that had the largest gains in the number of patents. The effect is quite substantial: a one-percentage-point increase in the number of foreign-born college graduates increased patents per capita by 9 to 18 percent. This large effect is explained by the fact that highly skilled immigrants directly contribute to innovation more than natives do, because they patent at twice the rate. In addition, by increasing local human capital, well-educated immigrants help states achieve the critical mass needed to sustain dynamic innovation hubs and foster knowledge spillovers. Thus immigration flows are further reinforcing the growing divide between American communities that has been taking place for the past three decades.

Until recently it was difficult to measure the contribution of different ethnicities to innovation, because the patent office does not track this. But in 2011 the economist William Kerr used an ethnic-name database similar to the ones used by marketers to measure exactly which ethnicities contribute the most to America's technological development. (Basically, if an inventor's family name is Chang, Kerr labeled her as Chinese; if the family name is Gupta, he labeled her as Indian.) He found a stunning increase during the 1990s in the share of patents granted to ethnic inventors, especially those of Chinese and Indian origin. Patents by ethnic inventors are concentrated in high-tech industries such as computers and pharmaceuticals, while the patents of natives are relatively more concentrated in traditional manufacturing areas.

All the existing studies show that immigrants account for a growing share of U.S. technology development. I am familiar with these statistics; I have been studying them for years. And yet nothing prepared me for what I saw when I actually visited high-tech companies in Silicon Valley. Anyone who doubts the role of immigrants in America's economy should go to one of these firms, have lunch in the company cafeteria, and speak with the employees. That is where you really understand how much of America's innovation is created by non-native-born Americans.

Jobs and Visas

During the 1990s, more than one million Soviet emigrants arrived in Israel, most of them highly educated. Given Israel's size, this amounted to an unprecedented increase in human capital. Although the impact on local manufacturing was disappointing, the high-tech sector experienced a significant jump in productivity and innovation. The same pattern emerges in other cases of mass migration of skilled individuals. On July 1, 1997, Great Britain handed over Hong Kong to China. Concerned about living under Chinese rule, thousands of Hong Kong residents, many of them wealthy and well educated, moved to Vancouver in the years preceding the handover. While there were some inevitable cultural tensions early on and not all the Chinese remained, in the end the city gained from this inflow in terms of both human and financial capital. The immigrants brought their savings, and the local economy received hundreds of millions of dollars in new investment. Many immigrants settled in towering condos reminiscent of the high-density high-rises back home, thus dramatically accelerating the revitalization of downtown. These changes helped turn Vancouver into a culturally diverse global metropolis.

Japan has had the opposite experience. Japanese high-tech companies dominated global markets in the 1980s, but they have lost much of their edge in the past twenty years, especially in software and Internet-related businesses. There are many explanations for this stunning reversal of fortunes, but a leading factor is that Japanese firms have access to a substantially smaller pool of software engineers than American firms, largely because of a lack of immigrants. While the United States is a magnet for the most talented foreign-born software engineers, legal, cultural, and linguistic barriers limit the inflow of global human capital into Japan and have effectively cost Japan its leadership in some of the most dynamic parts of the high-tech sector. As we saw earlier, the thickness of the labor market for specialized occupations is a crucial factor in determining the success of the innovation sector.

Today the contentious debate on immigration in America misses a key point: a visa issued to a highly skilled immigrant does not necessarily mean one less job for an American citizen. On the contrary, it could mean many more jobs for American citizens. While foreign-born workers account for 15 percent of America's labor force, they account for a third of all engineers and half of all those with doctorates. Without immigrants, the United States would not dominate the sciences the way it does now. For one thing, it would have many fewer Nobel Prizes. Foreign-born scientists working in the United States are more than twice as likely to win a Nobel Prize as their American-born colleagues, and they are also overrepresented among members of the National Academy of Sciences and the National Academy of Engineering. But it is not only about scientific awards and academic titles. Even more important for American workers, immigrants are almost 30 percent more likely than nonimmigrants to start a business, and they account for one-quarter of all venture-backed public companies since 1990 and one-quarter of new high-tech firms with over $1 million in sales. Steve Jobs (whose Syrian father came to the United States for his doctoral studies), Jerry Yang (the Taiwanese-born cofounder of Yahoo), and Sergey Brin (the Russian-born cofounder of Google) are just a few examples of immigrants or their children who created businesses that have gone on to provide thousands of new jobs for American natives.

Viewed this way, the current debate on immigration looks misguided. It has devolved into an ideological fight between those who want tougher rules and those who want softer rules. But the key question is not how many immigrants but *what kind* of immigrants to let in. It is probably the case that unskilled immigrants tend to depress the salaries of unskilled natives and therefore exacerbate inequality, although the exact magnitude of that effect is still the subject of lively debate among economists. But the effect of highly skilled immigrants is apt to be positive, especially for low-skilled Americans.

There are three reasons for this. First, high-skilled immigrants do not compete directly with low-skilled Americans. In fact, the two complement each other, which means that an increase in the former group is likely to raise the productivity of the latter group. Second, firms are apt to respond to an inflow of highly skilled immigrants by investing more, and this new investment may further raise the productivity of low-skilled workers. Third, skilled immigrants generate important spillovers at the local level, since an increase in the number of highly educated individuals in a city tends to strengthen the local economy, thus generating local jobs and raising natives' wages.

In principle, it is possible that the effect is positive even for skilled Americans. Obviously, highly skilled immigrants compete with their American counterparts, and this would tend to depress Americans' wages. But the other two effects would tend to push the skilled natives' wages in the other direction, potentially offsetting the negative effect. Regardless, a substantial increase in the number of skilled immigrants could play an important role in reducing wage inequality. Overall, limiting the number of unskilled immigrants is unlikely to have major negative effects for natives, but limiting the number of *skilled* immigrants could have significant negative effects, especially for our low-skilled workers.

Recent research by Jennifer Hunt identifies which kind of highly skilled immigrant is most likely to bring benefits to American natives. Using a detailed sample of college-educated immigrants, she found that those who arrived as postdoctoral fellows and medical residents have been most successful in generating original research and patents and vastly outperform natives. By contrast, immigrants who arrived thanks to a family member who was already in the United States perform at the same level as natives.

It is in America's self-interest to radically reform its immigration policy to favor immigrants with college degrees, master's degrees, and PhDs. Right now, 60 percent of the students in American engineering schools are foreign-born, but when these individuals graduate, they often find it difficult to stay in the United States. Several human resource managers of high-tech companies have told me that the current U.S. policy acts as a constraint on their ability to expand. One HR manager in Silicon Valley was very explicit: she called the current policy "ridiculous." It is costly and time-consuming for these companies to hire qualified high-tech employees who are born abroad, even if those individuals have a master's degree and graduated at the top of their class from Stanford or MIT. The number of visas for skilled workers, called H1B visas, is too low given the needs of high-tech companies, and during normal years such visas run out shortly after they are made available. An employee at Intel recently told me that the company, one of the largest users of H1B visas in the nation, often hires a law firm that in turn hires a paralegal to spend the night in front of the federal building where applications are submitted so that Intel's applications are among the first to be turned in. (Because of the weak labor market, the years 2009–2011 have been a significant exception, with many more slots than visa petitions.) A friend of mine (who obviously wants to remain anonymous) has an MBA from Berkeley and received a job offer from a leading high-tech firm but had to marry her American partner just to avoid visa delays and secure her job. We should try to do everything in our power to keep this kind of person, and instead we do everything we can to discourage her from staying. Our ability to absorb the world's talent is a crucial advantage no other culture can match. But it is constrained by an immigration policy that goes against our own economic interests.

America's Choice

At the beginning of the twentieth century, the United States was still a largely undeveloped country, with a brief history and a culture that was clearly inferior to that of Europe. Intellectually, Berlin, Paris, London, and Rome were the Western world's powerhouses. They considered New York and Chicago provincial outposts at best, while the rest of the country was untamed and unworthy of attention. Nevertheless, around that time America established itself as the undisputed leader in education. Unique among industrialized nations, it decided to make high school essentially universal; by contrast, European countries—traditionally more elitist in all matters—waited several decades. And when they did start educating all of their children, they were much less progressive and ambitious than the United States.

This farsighted decision meant that for most of the twentieth century, the United States remained the world leader in human capital investment. In their recent book, Harvard's Goldin and Katz call the twentieth century the "human capital century." The American worker was so much better educated than the workers of other countries that he became the most productive, innovative, and entrepreneurial in the world. Goldin and Katz note that it is not an accident that the human capital century was also the American century. America's unstoppable rise from unsophisticated out-post on the world's periphery to global economic superpower had a lot to do with the superior skills of its workforce. Through most of the twentieth century, the United States was the world leader in education. Its economic dominance was in no small part due to its educational dominance. But over the past thirty years, this aggressive policy of educational expansion has lost ground. While American graduate schools and research institutions remain the best in the world, the country's elementary and secondary schools lag behind those of many European countries and a growing number of developing countries, while college graduation rates have slowed down.

If human capital was the key to economic prosperity in the twentieth century, it is even more important in the twenty-first. In the coming decades, successful societies will be the ones that can attract and nurture the most creative workers and entrepreneurs. The United States needs to choose how it wants to increase its human capital to stay competitive in this new economy. There are two ways to supply America's innovative businesses with the educated workers they need while reducing the economic divide between those with skills and those without.

One avenue is to dramatically reform immigration policy in favor of workers with college and postgraduate degrees. This policy, already adopted by such countries as Canada and Australia, would increase America's human capital at little expense to American taxpayers. Allowing an Indian engineer to immigrate to the United States instead of educating an American engineer means that U.S. companies can draw on his talent without requiring U.S. taxpayers to pay for his education. The United States essentially receives free human capital, courtesy of India. The alternative is to increase human capital in the United States by educating Americans. This choice involves considerable costs for American taxpayers in the short run, since it would involve revamping high school education and significantly expanding higher education, but it also involves considerable long-run benefits for Americans who become better educated and end up with good jobs. Doing nothing is certainly an option, but it is a terrible one. It would mean losing our advantage in innovation. Worse, it would lead to stagnation and irreversible atrophy.

The choice between education and immigration is not neutral. To Google, it might not matter whether its engineers have an American passport or an Indian one, as long as it can pick the best and the brightest, but to American workers it is critical. It means that the substantial rewards of that high-tech job do not go to a U.S. worker but to an Indian one. We would be giving the innovative jobs to well-educated foreigners, leaving ourselves with the service jobs created by the multiplier effect. This is a world in which the iPhone is designed and engineered in Cupertino by Chinese or Indian PhDs and native American workers are the waiters, carpenters, and nurses who support them. What America decides about education is one of the most important strategic decisions facing the country today.

The Local Global Economy

As the great urban thinker Jane Jacobs recognized fifty years ago, communities, just like natural ecosystems, are not static entities but continually evolving creative commons that expand or shrink depending on the ingenuity of their residents. They are *human ecosystems*. Their process of continuous destruction and regeneration is ultimately what drives innovation, today as in the past. Innovation makes some activities and occupations obsolete while creating new ones. The creative spark is constantly generating what Jacobs called "new work." She was writing in the 1960s and 1970s, so her examples of innovation are old-fashioned, but her vision of what makes a society vital and prosperous still rings true today: innovation happens when people interact in a fertile urban environment and their ideas unexpectedly collide to create something that did not exist before.

We live in a world that is full of paradoxes. This makes it sometimes challenging to comprehend, but also incredibly fascinating. One of the most intriguing paradoxes is that our global economy is becoming increasingly local. Despite all the hype about exploding connectivity and the death of distance, where we live and work is more important than ever. Our best ideas still reflect the daily, unpredictable stimuli that we receive from the people we come across and our immediate social environment. Most of our crucial interactions are still face-to-face, and most of what we learn that is valuable comes from the people we know, not from Wikipedia. The vast majority of the world's phone calls, Web traffic, and investments are still local. Telecommuting is still incredibly rare. Videoconferencing, e-mail, and Skype have not made a dent in the need for innovative people to work side by side. In fact, that is more important than ever. At the same time that goods and information travel at faster and faster speeds to all corners of the globe, we are witnessing an inverse gravitational pull toward certain key urban centers. Globalization and localization seem to be two sides of the same coin. More than ever, local communities are the secret of economic success. As Yaniv Bensadon, an Israeli entrepreneur who recently moved his startup to Silicon Valley, put it, "It is true that things can be done anywhere on the Internet, but at the end of the day, it's still a people business."

These two major trends of the twenty-first century—increased globalization and increased localization—are reshaping our work environments and the very fabric of our communities. They are also redefining America's role in the world. Although we are no longer the dominant producer of material goods, we are striving to maintain our role as the dominant producer of knowledge and new ideas. To succeed, we need to regain our unity and refocus our priorities. Above all, we need to decide which America we want as our future—the America of ever-increasing educational levels, rising productivity, and pragmatic optimism, or the America of deteriorating skills, shrinking horizons, and paralyzing pessimism. We are at one of those major historical crossroads that determine the fate of nations for decades to come. Although we face serious challenges, we have strengths that no other society can match. Our unparalleled ability to attract and welcome the most creative individuals from all over the world, the dynamism of our workplaces, and the strength of our brain hubs give us a significant head start in this new global economy. It is up to us to keep it.

ACKNOWLEDGMENTS

Serious academic economists are not supposed to write books—they are supposed to write technical papers. Indeed, writing such papers has been my main concern for the past fifteen years. Economics as a field does not reward popular writing, unlike some other academic disciplines. There are many good reasons for this, but after spending fifteen years doing research on questions at the intersection of labor and urban economics, I developed an increasing desire to reach a larger audience than the one that reads my technical papers. A fateful conversation with Matt Kahn and a year of sabbatical were the precipitating factors that finally induced me to start a project that I had been considering for a few years. It turned out to be a surprisingly pleasant experience. In some serendipitous way, spending a year looking at the big picture gave me a lot of new ideas for future research, and I can't wait to start writing papers again.

One of the reasons this project turned out to be so interesting was that it gave me the chance to talk to many people who influenced my thinking. I wrote most of the text during my sabbatical at the Stanford economics department and the Stanford Institute for Economic Policy Research. I am grateful to John Shoven for his wonderful hospitality and for taking the time to connect me with some brilliant Silicon Valley innovators. Many friends and colleagues provided excellent suggestions. In particular, Chang-Tai Hsieh, Paul Oyer, Pat Kline, Giovanni Peri, Jesse Rothstein, Alex Mas, Michael Yarne, and Willem Vroegh read early drafts (or parts of early drafts) and were generous with insightful comments and constructive criticism. Bruce Mann and Matt Warning were kind enough to provide me with useful information about the history of Seattle. Throughout the project I benefited from conversations with Nick Bloom, John Van Reenen, Severin Borenstein, Ward Hanson, Mark Breedlove, Ed Glaeser, Jacques Lawarree, Gerald Autler, Marc Babsin, Ben Von Zastrow, Marco Tarchini, Ted Miguel, Antonio Moretti, and Giacomo De Giorgi. Alex Wolinsky and Joyce Liu, two exceedingly bright, curious, and hardworking Berkeley students, provided excellent research assistance and helped me improve the text. The skilled cartographer Mike Webster created the maps in Chapter 3.

My agent, Zoë Pagnamenta, expertly guided me through the process of finding a publisher. One of the reasons that I was attracted to Houghton Mifflin Harcourt was Amanda Cook, who was described to me as one of the best editors in the business. She did not disappoint. Her feedback helped me shape the text and finesse the concepts with both detailed comments and big-picture suggestions. My manuscript editor, Liz Duvall, further improved the text with two rounds of careful editing and many useful suggestions.

Above all, I am grateful to Ilaria. With unreasonable optimism, she has always supported the idea of trying something different, even when my unreasonable pessimism led me to postpone and delay. She has invariably been right.

NOTES

page **INTRODUCTION**

1 *In 1969, David Breedlove was a young engineer:* Personal conversation with Mark Breedlove, son of David Breedlove.

8 *Shenzhen's population has grown:* According to the Shenzhen government's official website, the city population was 30,000 in 1979 and 10.36 million in 2010.

11 *Essentially this is why:* Kraemer, Linden, and Dedrick, "Capturing Value in Global Networks."

12 *About a third of Americans work:* My calculations are based on County Business Patterns data from the U.S. Census Bureau.

14 *"the New Economy gives both companies and workers":* Atkinson and Gottlieb, "The Metropolitan New Economy Index," 2001.

1. AMERICAN RUST

19 *American Rust: American Rust* is the title of a novel by Philipp Meyer.

22 *Consider Figure 1:* My analysis is based on data from County Business Patterns, U.S. Census Bureau.
Nineteen out of twenty sectors: U.S. Bureau of Labor Statistics, "Industry Output and Employment Projections to 2018."

24 *"the United States no longer":* Jacoby, "Made in the USA."

27 *In the decade after World War II:* Glaeser, *Triumph of the City.*

28 *"People think China is cheap":* Fallows, "China Makes, the World Takes."

An important new study: Autor, Dorn, and Hanson, "The China Syndrome."

29 *A recent study by Nicholas Bloom:* Bloom, Draca, and Van Reenen, "Trade Induced Technical Change?"

31 *"It seems as if every 28-year-old guy":* Karrie Jacobs, "Made in Brooklyn," *Metropolis,* June 2010.
 "Most of them are under thirty": Ibid.

34 *Two economists at the University of Chicago:* Broda and Romalis, "The Welfare Implications of Rising Price Dispersion."

35 *The economist Emek Basker:* Basker, "Selling a Cheaper Mousetrap."

37 *Take a look at the evolution:* My analysis is based on data from County Business Patterns, U.S. Census Bureau.

39 *"Everyone in America has heard of Dell":* Fallows, "China Makes, the World Takes."

41 *It's hollowing out:* Autor, "The Polarization of Job Opportunities in the U.S. Labor Market."
 In an influential 2003 paper: Autor, Levy, and Murnane, "The Skill Content of Recent Technological Change."

42 *Autor has looked at changes in employment:* Autor, "The Polarization of Job Opportunities in the U.S. Labor Market."

2. SMART LABOR: MICROCHIPS, MOVIES, AND MULTIPLIERS

47 *The number of patents granted around the world:* World Intellectual Property Indicators, World Intellectual Property Organization, 2010.
 In the United States, the top patent producers: Data come from the author's analysis of data from the U.S. Patent and Trademark Office.

48 *"misplaced faith":* Grove, "How America Can Create Jobs."

49 *As you can see in Figure 3:* The author's analysis is based on data from County Business Patterns, U.S. Census Bureau.
 A recent study estimates: Hann, Viswanathan, and Koh, "The Facebook App Economy."

52 *The economist Michelle Alexopoulos has painstakingly assembled:* Alexopoulos, "Read All About It!!"
 Similarly, the global management consulting company McKinsey: Manyika and Roxburgh, "The Great Transformer."
 With an impressive 300 percent growth: According to the Bureau of Labor Statistics, "workers in this industry conduct much, but not all, of the scientific research and R&D in the economy . . . Much of the R&D conducted by companies in a wide range of industries—such as pharmaceuticals, chemicals, motor vehicles, and aerospace products—is conducted within the scientific research and development

services industry, because many companies maintain laboratories and other R&D facilities that are located apart from production plants and other establishments characteristic of these industries." From Bureau of Labor Statistics, "Career Guide to Industries, 2010–2011 Edition."
The Bureau of Labor Statistics puts biomedical engineers: Bureau of Labor Statistics, "Occupational Employment Projections to 2018."

53 *A recent study shows:* Balasubramanian and Sivadasan, "What Happens When Firms Patent?"
"tens of thousands of jobs": Calvey, "Bay Area Startups Court Cash-strapped, Creditworthy."

56 *This industry is generating:* "Yoga in America," *Yoga Journal*, February 2008.
Tens of thousands of people: Bureau of Labor Statistics, *Occupational Outlook Handbook,* 2010–2011.

60 *With only a fraction of the jobs:* Not all traded jobs are alike. Cities where workers in the traded sector are more productive and have higher incomes can support a larger non-traded sector. Data show that in the United States, cities with more and better-paying jobs in the traded sector also have more non-traded jobs.
My research, based on an analysis: Moretti, "Local Multipliers," and follow-up analysis by the author.

61 *My analysis indicates:* Ibid.
"If you get an auto assembly plant": Wessel, "The Factory Floor Has a Ceiling on Job Creation."
According to a company report: Eicher, "The Microsoft Economic Impact Study."

65 *An analysis of the French Internet sector:* Pélissié du Rausas et al., "Internet Matters."

66 *"Tech insiders thought":* Helft, "In Silicon Valley, Buying Companies for Their Engineers."

68 *American exports to China have increased:* Barboza, "As China Grows, So Does Its Appetite for American-Made Products."

69 *Let's consider Oracle:* Wessel, "Big U.S. Firms Shift Hiring Abroad."

70 *"making $1,000 a year":* Fallows, "China Makes, the World Takes."
India was about to receive certification: More precisely, at the time India was making efforts to be classified as fully adherent to a set of laboratory testing rules defined by the working group on Good Laboratory Practices of the OECD. India was accepted as a full member of the working group on GLP in OECD with full obligations on March 3, 2011. This implies that safety data generated in preclinical testing done in GLP-certified labs in India would now be acceptable in all OECD member countries, including all of Europe, the United States, and Japan, and also in nonmember countries such as South Africa and

Singapore. This would include testing of pharmaceuticals, industrial chemicals, food and feed additives, and agrochemicals, which may be synthetic or biological in origin.

A series of studies: Cockburn and Slaughter, "The Global Location of Biopharmaceutical Knowledge Activity"; Hanson, Mataloni, and Slaughter, "Expansion Abroad and the Domestic Operations of U.S. Multinational Firms."

71 *"Offshoring appears to have contributed":* National Academy of Engineering, *The Offshoring of Engineering.*

The economists Natarajan Balasubramanian and Jagadeesh Sivadasan: Balasubramanian and Sivadasan, "What Happens When Firms Patent?"

The London School of Economics professor: Van Reenen, "The Creation and Capture of Rents."

72 *Thus the economic rent:* The fact that workers capture some of the economic rent of innovative activity makes sense. After all, the most important ingredient in innovation is not physical capital but human capital. When firms invest in R&D, the majority of their spending is on salaries for R&D workers. While labs and machines are important, people are a much more important ingredient. Workers capture so much of the economic rent generated by innovation because jobs in innovation are different from other jobs—they involve creativity and ingenuity. It is difficult to inspire creativity using standard compensation packages. Because creativity and innovation require a much more personal involvement on the part of workers than standard, less creative jobs, rent sharing between workers and employers is a way for the company to urge its employees to work harder.

3. THE GREAT DIVERGENCE

76 *"city of despair":* "City of Despair," *The Economist,* May 22, 1971, pp. 57–58.

77 *Before the move, the labor markets:* The figures are based on my calculations using data from the Census of Population and the American Community Survey. Workers with a college education include those with a postgraduate degree.

80 *By one estimate, Microsoft alumni:* Tice, "Geeks of a Feather."

81 *I estimate that Microsoft is responsible:* The calculations are based on estimates in Moretti, "Local Multipliers."

83 *The accompanying map shows:* My analysis is based on data from the U.S. Patent and Trademark Office.

The states that generate the most patents: The calculations are based on data from the U.S. Patent and Trademark Office, "Extended Year

Set—Patents By Country, State, and Year All Patent Types," December 2010. Hunt and Gauthier-Loiselle document that in the years between 1929 and 1989 there was convergence in patenting across states, but divergence started in 1990. See Hunt and Gauthier-Loiselle, "How Much Does Immigration Boost Innovation?"

85 *It accounts for more than a third:* Data are from PricewaterhouseCoopers, "MoneyTree Report." The earliest available year is 1995.
Although they are not particularly close: Echeverri-Carroll, "Economic Growth and Linkage with Silicon Valley."

89 *Using data collected by the Census Bureau:* I report the weighted average of yearly or hourly labor earnings (before taxes) from the American Community Survey in 2006, 2007, and 2008, obtained from Ipums USA. To maximize the precision of my estimates, I include in the analysis of earnings by occupation only cities for which there are at least one hundred observations in the relevant occupation. For example, for lawyers, I only consider cities for which the American Community Survey reports earnings data for at least one hundred lawyers. In the case of industrial production managers, I include in the analysis cities for which there are at least twenty observations in the relevant occupation, because there are too few cities with one hundred valid observations.

90 *Map 2 shows just how different:* My analysis is based on data from the American Community Survey, U.S. Census Bureau.

92 *Table 1 shows:* Tables 1 and 2 are based on my calculations using data from the American Community Survey, U.S. Census Bureau, for 2006–2008. The sample includes all workers aged twenty-five through fifty-nine with nonmissing information on education and earnings who live in large metropolitan areas (working population above 200,000) that are identified in all relevant years. Earnings are defined as yearly labor earnings before taxes. Weighted averages are displayed, where the weights are the relevant "person weights" provided by the ACS.

97 *Figure 4 shows the relationship:* My analysis is based on data from the American Community Survey, U.S. Census Bureau. All metropolitan areas are included.

99 *This relationship holds for all sectors:* Moretti, "Estimating the Social Return to Higher Education"; Moretti, "Workers' Education, Spillovers and Productivity."

100 *In a famous 1988 article:* Lucas, "On the Mechanics of Economic Development."
In a study that I published in 2004: Moretti, "Estimating the Social Return to Higher Education."
The economist Jeffrey Lin: Lin, "Technological Adaptation, Cities, and New Work."

102 *Map 3 shows the change:* My analysis is based on data from the Cen-

sus of Population and the American Community Survey, U.S. Census Bureau.

A good way to appreciate the Great Divergence: My calculations are based on microdata from the Censuses of Population for 1980, 1990, and 2000 and the American Community Survey for 2006, 2007, and 2008, from the U.S. Census Bureau. Workers aged twenty-five to sixty are included. The estimates are based on the year-specific definition of top ten and bottom ten metropolitan areas. Thus the metro areas in the top and bottom groups vary from year to year. The trends in the graph are therefore to be interpreted as changes in cross-sectional inequality. Alternatively, one could define the top and bottom groups based on college share in 1980 (that is, the ten cities with the most and the least college share in 1980). In that case, the identity of the cities in each group would be fixed. The picture would be qualitatively similar, as can be seen below:

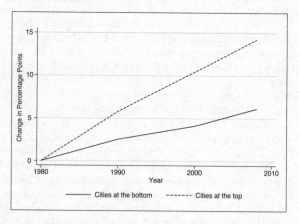

Figure 12: Gains in the share of college graduates since 1980 (groups based on 1980 college share)

Evidence indicates that American cities: Glaeser and Vigdor, "The End of the Segregated Century."

105 *Indeed, Figure 6 shows the gains:* Like the previous figure, this figure is based on my calculations using microdata from the Censuses of Population for 1980, 1990, and 2000 and the American Community Survey for 2006, 2007, and 2008, both from the U.S. Census Bureau. Workers aged twenty-five to sixty are included. As before, these estimates are based on the year-specific definition of the top ten and bottom ten metropolitan areas (in other words, the figure shows changes in

the cross-sectional inequality). Thus, the metro areas in the top and bottom groups vary in each year. If the top and bottom groups were defined based on college share in 1980 (that is, the ten cities with the most and the least college share in 1980), the picture would be qualitatively similar, as seen below:

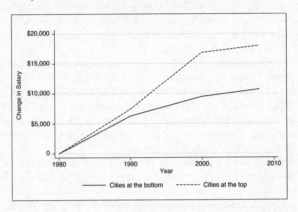

Figure 13. Gains in earnings of college graduates since 1980 (groups based on 1980 college share)

107 *Map 4 shows just how different:* My analysis is based on data from the Institute for Health Metrics and Evaluation, University of Washington.
109 *This degree of geographical inequality:* Kulkarni, Levin-Rector, Ezzati, and Murray, "Falling Behind."
The rising inequality in life expectancy: My calculations are based on life expectancy data for 3,147 U.S. counties for men and women from the Institute for Health Metrics and Evaluation, University of Washington. As for the figures that show divergence in education and earnings, these estimates are based on the year-specific definition of the top ten and bottom ten counties. Thus the counties in the top and bottom groups vary from year to year.
111 *The economist and former air force officer:* Carrell, Hoekstra, and West, "Is Poor Fitness Contagious?"
112 *The Yale economist Jason Fletcher:* Fletcher, "Social Interactions and Smoking."
The Moving to Opportunity program: Ludwig et al., "Neighborhoods, Obesity, and Diabetes—A Randomized Social Experiment."
113 *Using data on 8 million adults:* The data are from the American Community Survey and include all individuals eighteen to seventy years of age who have ever been married.

114 *Figure 8 shows the increase:* My analysis is based on data from the Census of Population and the American Community Survey, U.S. Census Bureau.

115 *In the 2008 presidential election:* All the figures in this section are based on my analysis of county-level data on votes cast. The data are the ones I used in my papers "Does Voting Technology Affect Election Outcomes?" and "Racial Bias in the 2008 Presidential Election."

116 *In research published in 2004:* Milligan, Moretti, and Oreopoulos, "Does Education Improve Citizenship?"

"a stable and democratic society": Friedman, *Capitalism and Freedom.*

Figure 9 shows changes in voter participation: My analysis is based on data from Dave Leip's Atlas of U.S. Presidential Elections and CNN.

119 *Which American cities have the most charities:* My calculations are based on 501c(3) organizations filing long forms in the IRS Statistics of Income data files. These data are the ones used in Card, Hallock, and Moretti, "The Geography of Giving."

difference between communities: The difference between the ten cities that had the highest per capita number of not-for-profit organizations (or the highest per capita contributions) and the ten cities that had the lowest per capita number of not-for-profit organizations (or the lowest per capita contributions) is much larger today than in 1990.

4. FORCES OF ATTRACTION

125 *"in search of [funding]":* Hoge, "Help Desk Firm Solves Problem of How to Grow."

"I knew I wanted to": Tam, "Technology Companies Look Beyond Region for New Hires."

127 *"The speed at which things move":* Kane, "Overseas Start-Ups Move In."

128 *For example, studies have shown:* Baumgardner, "Physicians' Services and the Division of Labor Across Local Markets."

Think about the history of Facebook: In a recent interview, Zuckerberg criticized several aspects of Silicon Valley's culture that he does not like but admitted that "Facebook would not have worked if I had stayed in Boston." Interview at Y Combinator's Startup School, October 29, 2011.

The size of labor markets: Wheeler, "Local Market Scale and the Pattern of Job Changes Among Young Men"; Bleakley and Lin, "Thick-Market Effects and Churning in the Labor Market."

130 *In a recent study of changing family structure:* Costa and Kahn, "Power Couples."

134 *"where Ericsson has more than 1,200 employees":* Clark, "Overseas Tech Firms Ramp Up Hiring in Silicon Valley."

136 *One study finds that the likelihood:* Sorenson and Stuart, "Syndication Networks and the Spatial Distribution of Venture Capital Investment." *"to be closer":* Delo, "When the Car-Rental Fleet Is Parked in Your Driveway." *"it is tough to get funding":* Gelles, "All Roads Lead to the Valley."

137 *"There's a lot of support":* Interview, "The Changing Role of the Venture Capitalist," *Marketplace*, NPR, January 18, 2011.

138 *"to be close to the action":* Kissack, "Electric Vehicle Companies Tap Silicon Valley Cash."

139 *"Knowledge flows are invisible":* Quoted in Jaffe, Trajtenberg, and Henderson, "Geographic Localization of Knowledge Spillovers as Evidenced by Patent Citations." *In 1993 three economists:* Ibid.

140 *Excluding citations that come from the same company:* Thompson, "Patent Citations and the Geography of Knowledge Spillovers." *"cricket spills over":* Lohr, "Silicon Valley Shaped by Technology and Traffic." *Citations are highest:* Belenzon and Schankerman, "Spreading the Word." *Geographical distance seems to impede:* Adams and Jaffe, "Bounding the Effects of R&D."

141 *Pierre Azoulay, Joshua Graff Zivin, and Jialan Wang:* Azoulay, Graff Zivin, and Wang, "Superstar Extinction."

142 *When a team of Harvard Medical School doctors:* Lee, Brownstein, Mills, and Kohane, "Does Collocation Inform the Impact of Collaboration?"

143 *The goal is "radical collaboration":* This definition is taken from the website of Hub SoMa, a social enterprise that is part of the Chronicle Building project.

147 *In 1993 the urban planner Ann Markusen:* Markusen, Hall, Campbell, and Deitrick, *The Rise of the Gun Belt*, pp. 187–89.

148 *The Princeton economist Alan Blinder:* Quoted in Grove, "How America Can Create Jobs."

149 *The economist Steven Klepper has shown:* Klepper, "The Origin and Growth of Industry Clusters."

151 *"turn on your TV":* Lindsay Riddell, "'Anti-Growth' Company Shooting for the Stars," *San Francisco Business Times*, July 1, 2011.

152 *Vacancies are so common:* Mattioli, "As Kodak Fades, Rochester Develops Other Businesses."

5. THE INEQUALITY OF MOBILITY AND COST OF LIVING

154 *In a study published in 2006:* Manacorda and Moretti, "Why Do Most Italian Youths Live with Their Parents?"

155 *"millions of men"*: Quoted in Ferrie, "Internal Migration."
"facilitated the exploitation": Ibid.
About 33 percent of Americans: Micro data, 2000 Census.

157 *Using data from millions of individual histories:* Wozniak, "Are College Graduates More Responsive to Distant Labor Market Opportunities?"

158 *In the United Kingdom:* Gregg, Machin, and Manning, "Mobility and Joblessness."
When Europeans are asked: Machin, Pelkonen, and Salvanes, "Education and Mobility."

159 *Figure 10 shows the difference:* Data from the U.S. Bureau of Labor Statistics. Data are based on individuals twenty-five years and older.

162 *Remarkably, this policy would also help:* In economic jargon, the market failure here is represented by the frictions in the process of job search that cause unemployment in the first place.

163 *"spatial mismatch":* Kain, "Housing Segregation, Negro Employment, and Metropolitan Decentralization."

164 *A team of University of Michigan economists:* Bound, Groen, Kézdi, and Turner, "Trade in University Training."

165 *The city of Norilsk:* Mallaby, *More Money Than God.*

167 *One study found that the larger the decline:* Sieg, Smith, Banzhaf, and Walsh, "Estimating the General Equilibrium Benefits of Large Changes in Spatially Delineated Public Goods."
To create the table, I used data: The data and the methodology are described in detail in my paper "Real Wage Inequality."

171 *However, this is not the end of the story:* I discuss these themes in some depth in "Local Labor Markets."

172 *When economists started measuring inequality:* See, for example, Krueger, Perri, Pistaferri, and Violante, "Cross-Sectional Facts for Macroeconomists."
In recent research, I found: Moretti, "Real Wage Inequality."
Therefore the difference in living standards: You might think, "All these college graduates may be paying sky-high mortgages and rents, but at least they live in great cities. The higher cost of living may simply reflect the better amenities in those cities." My research indicates that a sudden love for urban living did not drive the migration of college graduates to expensive coastal cities. While urban living has certainly become better than it was in the 1980s, that is more effect than cause. The main reason college graduates have moved to expensive urban areas has more to do with jobs. Jobs for college graduates, especially in finance and high tech, have increased more in expensive coastal cities, which brings graduates to those cities. College graduates are paying a high price to live in places like San Francisco and Boston, because their jobs are there, not because in 1980 they started liking San Francisco and Boston more than high school graduates did.

176 *a series of recent studies:* Glaeser and Ward, "The Causes and Consequences of Land Use Regulation"; Glaeser and Tobio, "The Rise of
the Sunbelt"; Glaeser, Gyourko, and Saks, "Why Is Manhattan So Expensive?"

6. POVERTY TRAPS AND SEXY CITIES

179 *Almost immediately, dozens of private biotech labs:* Powell, Whittington,
and Packalen, "Organizational and Institutional Genesis."
"Finding great science": Leuty, "SF Life Science Hub Lures East Coast
Venture Firms."
"Torrey Pines Road": Powell, Whittington, and Packalen, "Organizational and Institutional Genesis," pp. 4–5.
"public uproar": Ibid.

180 *Is it a question:* Zucker, Darby, and Brewer, "Intellectual Human Capital and the Birth of U.S. Biotechnology Enterprises."
When biotech appeared: Ibid.; Powell, Whittington, and Packalen, "Organizational and Institutional Genesis." In particular, Powell, Whittington, and Packalen argue that numerous potential clusters, each
with abundant endowments, could have evolved into important clusters, but they never did: "The New York City metropolitan area and
central New Jersey are both home to leading universities, . . . many
wealthy financial institutions, and numerous large multinational pharmaceutical companies . . . The Philadelphia metropolitan area had the
University of Pennsylvania, the Wistar Institute, the Fox Chase Can-
cer Center, and the Children's Hospital of Philadelphia, all important
public research organizations, and a number of major pharmaceutical companies as well . . . Los Angeles, where one of the earliest and
most successful biotech companies, Amgen, was founded in 1980, had
ample scientific resources at CalTech and UCLA, but a cluster never
cohered there. Houston, Texas, had financial wealth, several medical
schools and universities, and a pathbreaking research hospital . . . Atlanta, Georgia, has the Centers for Disease Control, research universities Emory and Georgia Tech . . . Cleveland, Ohio, was an early home
to venture capital, and the Cleveland Clinic is one of the premier research hospitals in the nation. Neither city today has significant activity in biotech" (pp. 4–5).

181 *In a fascinating and now classic study:* Zucker, Darby, and Brewer,
"Intellectual Human Capital and the Birth of U.S. Biotechnology
Enterprises"; Zucker et al., "Minerva Unbound"; Zucker, Darby,
and Armstrong, "Commercializing Knowledge"; Zucker and Darby,
"Capturing Technological Opportunity via Japan's Star Scientists";
Zucker, Darby, and Armstrong, "Geographically Localized Knowl-

edge"; Zucker and Darby, "Present at the Biotechnological Revolution."

182 *The attractive nature of economic development:* In more recent work, Zucker and Darby show that today the United States has just over half of the world's biotech stars. Even more interesting is how these numbers have been changing over time. The United States, with a strong lead in research universities, is the primary producer of star scientists in the world, although not all of the stars decide to stay. On net, however, stars show a clear tendency to concentrate in the United States. Recently stars have come to America from countries such as Switzerland, the United Kingdom, and Canada. This probably reflects the facts that new restrictions by Swiss cantons have made them inhospitable to biotechnology; the United Kingdom has systematically reduced university support; and the U.S. labor market for highly skilled individuals remains relatively more attractive than the Canadian one. Overall, the United States has managed to achieve positive net gains and therefore to increase its lead in the biotech sector.

The importance of stars to the locations of innovation clusters is evident in most high-tech industries, with the effect largest for pharmaceutical/biotech and computing/IT, smaller in nanotechnology, and smallest in semiconductors. See Zucker and Darby, "Movement of Star Scientists and Engineers and High-Tech Firm Entry."

183 *In 1913, the year before World War I began:* All the figures are from Scott, "Origins and Growth of the Hollywood Motion-Picture Industry."

184 *In 2006, the UCLA geographer Allen Scott proposed:* Ibid.

188 *"Seattle was the home":* Richard Florida, *The Rise of the Creative Class,* p. 206.

189 *"The arts have become":* Jacobs, "Made in Brooklyn."

192 *According to one study:* See "The Cost of Cool," *Economist,* September 17, 2011.

193 *"Our location has helped us":* Jon Swartz, "San Francisco's Charm Lures High-Tech Workers," *USA Today,* December 6, 2010.
   "We're able to attract": Ibid.

194 *"I try to find partners":* "The Revolution on Batteries," *Boston Globe,* May 22, 2011.

196 *A study by Adam Jaffe:* Jaffe, "Real Effects of Academic Research."

200 *Between 1933 and 1958:* The $30 billion are measured in 2010 dollars.
   "in practice, they work miserably": Jacobs, "Why TVA Failed."

201 *My colleague Pat Kline and I:* Kline and Moretti, "Local Economic Development, Agglomeration Economies and the Big Push."

202 *Consider Portland:* These three cases are described in detail in Mayer, "Bootstrapping High-Tech."

203 *As pointed out in a recent Brookings Institution study:* Ibid. Mayer adds that public policy has become more important in more recent years.

Local companies started advocating for public policy changes, and in some cases they were successful at engaging civic leaders. For example, as innovative companies gained size in Portland, Boise, and Kansas City, they began to seek more skilled workers and became more dependent on university-based research. In all cases, though, policy changes took time; technology-based metropolitan economic development is necessarily a long-term effort.

204 *"You can't throw a Frisbee"*: Vara, "Clean Tech Arrives, with Limited Payoff."

206 *At the beginning of the last decade:* Worldwide, patent applications for innovations in fuel cells and solar, wind, and geothermal energy increased almost fivefold in nine years, from 584 in 2000 to 3,424 in 2009. Solar energy accounted for the lion's share of this increase, with Japanese inventors generating the largest number of patents for solar energy and fuel cell technology and American inventors generating the largest number of patents for wind technology. See World Intellectual Property Organization, *World Intellectual Property Indicators,* 2010.
While there has been growth: Vara, "Red Flags for Green Energy."
As noted by my colleague: Borenstein, "The Private and Public Economics of Renewable Electricity Generation."

207 *Many American companies keep their headquarters:* Baker, "U.S. Solar Firms Lead in Installation."

208 *There are now twenty-five clean-tech firms:* Vara, "Clean Tech Arrives."

209 *With two colleagues, I have studied:* Greenstone, Hornbeck, and Moretti, "Identifying Agglomeration Spillovers"; Greenstone and Moretti, "Bidding for Industrial Plants."

211 *In 2010 a team of three economists:* Busso, Gregory, and Kline, "Assessing the Incidence and Efficiency of a Prominent Place Based Policy."

213 *Assuming that most of the jobs:* I am basing these calculations on the most recent version of the paper available at the time of writing (2011). The return is calculated by comparing the total annual increase in earnings for residents and nonresidents with the total cost of the program for federal taxpayers ($600 million) inflated by a factor equal to 1.3 to account for the efficiency losses associated with taxation. This calculation assumes that the shadow price of the labor of workers affected by the program is zero. Given that the program targeted high unemployment areas, this assumption is not unrealistic.

7. THE NEW "HUMAN CAPITAL CENTURY"

219 *In one of the most rigorous studies:* Bloom, Schankerman, and Van Reenen, "Identifying Technology Spillovers and Product Market Rivalry." See also Griffith, Harrison, and Van Reenen, "How Special Is

the Special Relationship?" and Lychagin, Pinske, Slade, and Van Reenen, "Spillovers in Space."

220 *Bloom and Van Reenen estimate:* Bloom, Schankerman, and Van Reenen,
"Identifying Technology Spillovers and Product Market Rivalry."

222 *Table 4 shows how:* My calculations are based on data from the Census of Population and the American Community Survey, U.S. Census
Bureau.

223 *If we dropped all the CEOs:* Indeed, the earnings data from the census that I used here are top-coded, which means that earnings above
$300,000 are set equal to $300,000. This is done to preserve the con-
fidentiality of the survey respondents. In the absence of top-coding, it
would be easy to identify very high earners in the data. A side effect
of top-coding is that very high salaries are not responsible for these
numbers.

224 *In the United States, wage inequality has increased:* See, for example, Au-
tor, "The Polarization of Job Opportunities in the U.S. Labor Market."
In a remarkable book, the Harvard economists: Goldin and Katz, *The Race
Between Education and Technology.*

226 *In 2011 the up-front cost:* Greenstone and Looney, "Where Is the Best
Place to Invest $102,000?"
The gap is large at age twenty-two: Ibid. These averages include both
those who have a job and those who do not have a job and have no salary, so they reflect not just earnings differences but also different unemployment probabilities.

228 *When the researchers Michael Greenstone:* Ibid.
In work based on a representative sample: Currie and Moretti, "Mother's
Education and the Intergenerational Transmission of Human Capital."

229 *In work with Lance Lochner:* Lochner and Moretti, "The Effect of Education on Crime."

230 *"Skill begets skill":* Heckman, "Policies to Foster Human Capital."

235 *"The decline in high school graduation":* Heckman and LaFontaine, "The
Declining American High School Graduation Rate."
"We cannot find": Samuelson, "High-Skill Job Openings Abound."
In one recent survey: Riddell, "East Bay Schools Aim to Refresh Biotech
Labor Pool."

237 *However, immigrants are also:* For example, Goldin and Katz, *The Race
Between Education and Technology,* report that 17 percent of immigrants
have been in school less than nine years, compared to only one percent
of natives.

238 *In 2010, Jennifer Hunt and Marjolaine Gauthier-Loiselle published:*
Hunt and Gauthier-Loiselle, "How Much Does Immigration Boost
Innovation?"

240 *But in 2011 the economist William Kerr:* Kerr, "The Agglomeration of
U.S. Ethnic Inventors."

241 *During the 1990s, more than one million:* Estimates for traditional manufacturing point to the opposite effect. See Paserman, "Do High-Skill
Immigrants Raise Productivity?"
Japan has had the opposite experience: Arora, Bransetter, and Drev, "Going Soft."

242 *Even more important for American workers:* All the figures are from Kerr,
"The Agglomeration of U.S. Ethnic Inventors."

244 *Recent research by Jennifer Hunt:* Hunt, "Which Immigrants Are Most
Innovative and Entrepreneurial?"

248 *"It is true that things":* Lohr, "Silicon Valley Shaped by Technology and
Traffic."

REFERENCES

Adams, James D., and Adam B. Jaffe. "Bounding the Effects of R&D: An Investigation Using Matched Establishment-Firm Data." *Rand Journal of Economics* 27, no. 4 (Winter 1996): 700–21.

Alexopoulos, Michelle. "Read All About It!! What Happens Following a Technology Shock?" *American Economic Review* 101, no. 4 (June 2011): 1144–79.

Arora, Ashish, Lee G. Bransetter, and Matej Drev. "Going Soft: How the Rise of Software Based Innovation Led to the Decline of Japan's IT Industry and the Resurgence of Silicon Valley." Working Paper 16156. National Bureau of Economics, July 2010.

Atkinson, Robert D., and Paul D. Gottlieb. "The Metropolitan New Economy Index." Progressive Policy Institute, Case Western Reserve University, 2001.

Autor, David. "The Polarization of Job Opportunities in the U.S. Labor Market: Implications for Employment and Earnings." Center for American Progress and the Hamilton Project, May 2010.

Autor, David H., and David Dorn. "The Growth of Low-Skill Service Jobs and the Polarization of the U.S. Labor Market." NBER Working Paper 15150. National Bureau of Economic Research, July 2009.

Autor, David H., David Dorn, and Gordon Hanson. "The China Syndrome: Local Labor Market Impacts of Import Competition in the United States." March 2011.

Autor, David H., Lawrence F. Katz, and Melissa S. Kearney. "The Polarization of the U.S. Labor Market." *American Economic Review* 96, no. 2 (May 2006): 189–94.

Autor, David H., Frank Levy, and Richard J. Murnane. "The Skill Content of Recent Technological Change: An Empirical Exploration." *Quarterly Journal of Economics* 118, no. 4 (November 2003): 1279–1334.

Azoulay, Pierre, Joshua S. Graff Zivin, and Jialan Wang. "Superstar Extinction." NBER Working Paper 14577. National Bureau of Economic Research, December 2008.

Baker, David R. "U.S. Solar Firms Lead in Installation." *San Francisco Chronicle*, September 24, 2011.

Balasubramanian, Natarajan, and Jagadeesh Sivadasan. "What Happens When Firms Patent? New Evidence from U.S. Manufacturing Census Data." *Review of Economics and Statistics* 93, no. 1 (February 2011): 126–46.

Barboza, David. "As China Grows, So Does Its Appetite for American-Made Products." *New York Times*, April 7, 2011.

Basker, Emek. "Selling a Cheaper Mousetrap: Wal-Mart's Effect on Retail Prices." *Journal of Urban Economics* 58, no. 2 (September 2005): 203–29.

Baumgardner, James R. "The Division of Labor, Local Markets, and Worker Organization." *Journal of Political Economy* 96, no. 3 (June 1988): 509–27.

———. "Physicians' Services and the Division of Labor Across Local Markets." *Journal of Political Economy* 96, no. 5 (October 1988): 948–82.

Becker, Gary S. *Human Capital: A Theoretical and Empirical Analysis, with Special Reference to Education.* New York: Columbia University Press, 1964.

Belenzon, Sharon, and Mark Schankerman. "Spreading the Word: Geography, Policy and University Knowledge Diffusion." CEPR Discussion Paper 8002. Centre for Economic Policy Research, September 2010.

Bishop, Bill. *The Big Sort: Why the Clustering of Like-Minded America Is Tearing Us Apart.* Boston: Houghton Mifflin Harcourt, 2008.

Bleakley, Hoyt, and Jeffrey Lin. "Thick-Market Effects and Churning in the Labor Market: Evidence from U.S. Cities." FRB of Philadelphia Working Paper 07-23. Philadelphia: Federal Reserve Bank, October 2007.

Bloom, Nicholas, Mirko Draca, and John Van Reenen. "Trade Induced Technical Change?: The Impact of Chinese Imports on Innovation, Diffusion IT and Productivity." NBER Working Paper 16717. National Bureau of Economic Research, January 2011.

Bloom, Nicholas, Mark Schankerman, and John Van Reenen. "Identifying Technology Spillovers and Product Market Rivalry." NBER Working Paper 13060. National Bureau of Economic Research, April 2007.

Borenstein, Severin. "The Private and Public Economics of Renewable Electricity Generation." NBER Working Paper 17695. National Bureau of Economic Research, December 2011.

Bound, John, Jeffrey Groen, Gábor Kézdi, and Sarah Turner. "Trade in University Training: Cross-State Variation in the Production and Use of College-Educated Labor." *Journal of Econometrics* 121 (2004): 143–73.

Broda, Christian, and John Romalis. "The Welfare Implications of Rising Price Dispersion." Chicago: University of Chicago, 2010.

Busso, Matias, Jesse Gregory, and Patrick M. Kline. "Assessing the Incidence and Efficiency of a Prominent Place Based Policy." NBER Working Paper 16096. National Bureau of Economic Research, June 2010.

Calvey, Mark. "Bay Area Startups Court Cash-strapped, Creditworthy." *San Francisco Business Times*, February 10, 2012.

Card, David, Kevin F. Hallock, and Enrico Moretti. "The Geography of Giving: The Effect of Corporate Headquarters on Local Charities." *Journal of Public Economics* 94, no. 3–4 (2010): 222–34.

Card, David, and Enrico Moretti. "Does Voting Technology Affect Election Outcomes? Touch-screen Voting and the 2004 Presidential Elections." *Review of Economics and Statistics* 89, no. 4 (November 2007): 660–73.

Carrell, Scott, Mark Hoekstra, and James E. West. "Is Poor Fitness Contagious? Evidence from Randomly Assigned Friends." *Journal of Public Economics* 95, no. 7–8 (August 2011): 657–63.

Clark, Don. "Overseas Tech Firms Ramp Up Hiring in Silicon Valley." *Wall Street Journal*, June 23, 2011.

Cockburn, Iain M., and Matthew J. Slaughter. "The Global Location of Biopharmaceutical Knowledge Activity: New Findings, New Questions." In Josh Lerner and Scott Stern, eds., *Innovation Policy and the Economy*, Vol. 10, pp. 129–57. Chicago: University of Chicago Press, 2010.

Costa, Dora L., and Matthew E. Kahn. "Power Couples: Changes in the Locational Choice of the College-Educated, 1940–1990." *Quarterly Journal of Economics* 115, no. 4 (November 2000): 1287–1315.

Currie, Janet, and Enrico Moretti. "Mother's Education and the Intergenerational Transmission of Human Capital: Evidence from College Openings." *Quarterly Journal of Economics* 118, no. 4 (2003): 1495–1532.

Delo, Stacey B. "When the Car-Rental Fleet Is Parked in Your Driveway." *Wall Street Journal*, May 19, 2011.

Echeverri-Carroll, Elsie. "Economic Growth and Linkage with Silicon Valley." *Texas Business Review*, December 2004.

Eicher, Theo. "The Microsoft Economic Impact Study." Seattle: Microsoft Corporation, March 2010.

Fallows, James. "China Makes, the World Takes." *The Atlantic*, July 2007.

Ferrie, Joe. "Internal Migration." In Susan B. Carter et al., eds., *Historical Statistics of the United States: Millennial Edition*. New York: Cambridge University Press, 2003.

Fletcher, Jason M. "Social Interactions and Smoking: Evidence Using Multiple Student Cohorts, Instrumental Variables, and School Fixed Effects." *Health Economics* 19, no. 4 (April 2010): 466–84.

Florida, Richard. *The Rise of the Creative Class: And How It's Transforming Work, Leisure, Community and Everyday Life.* New York: Basic, 2002.

Friedman, Milton. *Capitalism and Freedom.* Chicago: University of Chicago Press, 1962.

Gelles, David. "All Roads Lead to the Valley." *Financial Times,* May 18, 2009.

Glaeser, Edward, and Jacob Vigdor. "The End of the Segregated Century: Racial Separation in America's Neighborhoods, 1890–2010." Civic Report 66. Manhattan Institute for Policy Research, January 2012.

Glaeser, Edward L. *Triumph of the City: How Our Greatest Invention Makes Us Richer, Smarter, Greener, Healthier, and Happier.* New York: Penguin, 2011.

Glaeser, Edward L., Joseph Gyourko, and Raven Saks. "Why Is Manhattan So Expensive? Regulation and the Rise in House Prices." *Journal of Law and Economics* 48, no. 2 (October 2005): 331–69.

Glaeser, Edward L., and Kristina Tobio. "The Rise of the Sunbelt." NBER Working Paper 13071. National Bureau of Economic Research, April 2007.

Glaeser, Edward L., and Bryce A. Ward. "The Causes and Consequences of Land Use Regulation: Evidence from Greater Boston." *Journal of Urban Economics* 65 (2009): 265–78.

Goldin, Claudia, and Lawrence F. Katz. *The Race Between Education and Technology.* Cambridge, Mass.: Belknap/Harvard University Press, 2008.

Greenstone, Michael, Rick Hornbeck, and Enrico Moretti. "Identifying Agglomeration Spillovers: Evidence from Winners and Losers of Large Plant Openings." *Journal of Political Economy* 118, no. 3 (2010): 536–98.

Greenstone, Michael, and Adam Looney. "Where Is the Best Place to Invest $102,000—In Stocks, Bonds, or a College Degree?" Hamilton Project, Brookings Institution, June 2011.

Greenstone, Michael, and Enrico Moretti. "Bidding for Industrial Plants: Does Winning a 'Million Dollar Plant' Increase Welfare?" NBER Working Paper 9844. National Bureau of Economic Research, July 2003.

Gregg, Paul, and Stephen Machin. "Child Development and Success or Failure in the Youth Labor Market." In David G. Blanchflower and Richard B. Freeman, eds., *Youth Employment and Joblessness in Advanced Countries,* pp. 247–88. Chicago: University of Chicago Press, 2000.

Gregg, Paul, Stephen Machin, and Alan Manning. "Mobility and Joblessness." In Richard Card, Richard Blundell, and Richard B. Freeman, eds., *Seeking a Premier Economy: The Economic Effects of British Economic Reforms, 1980–2000,* pp. 371–410. Cambridge, Mass.: NBER Books, 2004.

Griffith, Rachel, Rupert Harrison, and John Van Reenen. "How Special Is the Special Relationship?" *American Economic Review* 96, no. 5 (2006).

Grove, Andy. "How America Can Create Jobs." *Bloomberg Businessweek*, July 1, 2010.

Hall, Matthew, Audrey Singer, Gordon F. De Jong, and Deborah Roempke Graefe. "The Geography of Immigrant Skills: Educational Profiles of Metropolitan Areas." *State of Metropolitan America* 34. Washington, D.C.: Brookings Institution, 2011.

Hann, Il-Horn, Siva Viswanathan, and Byungwan Koh. "The Facebook App Economy." Center for Digital Innovation, Technology and Strategy, University of Maryland, September 11, 2011.

Hanson, Gordon H., Jr., Raymond J. Mataloni, and Matthew J. Slaughter. "Expansion Abroad and the Domestic Operations of U.S. Multinational Firms." 2003.

Heckman, James. "Policies to Foster Human Capital." *Research in Economics* 54 (2000): 3–56.

Heckman, James J., and Paul A. LaFontaine. "The Declining American High School Graduation Rate: Evidence, Sources, and Consequences." *NBER Reporter*, 2008, no. 1.

Helft, Miguel. "In Silicon Valley, Buying Companies for Their Engineers." *New York Times*, May 18, 2011.

Hoge, Patrick. "Help Desk Firm Solves Problem of How to Grow." *San Francisco Business Times*, October 2010.

Hunt, Jennifer. "Which Immigrants Are Most Innovative and Entrepreneurial? Distinctions by Entry Visa." *Journal of Labor Economics* 29, no. 3 (July 2011): 417–57.

Hunt, Jennifer, and Marjolaine Gauthier-Loiselle. "How Much Does Immigration Boost Innovation?" *American Economic Journal: Macroeconomics* 2, no. 2 (April 2010): 31–56.

Jacobs, Jane. "Why TVA Failed." *New York Review of Books*, May 10, 1984.

Jacobs, Karrie. "Made in Brooklyn." *Metropolis*, June 2010.

Jacoby, Jeff. "Made in the USA." *Boston Globe*, February 6, 2011.

Jaffe, Adam. "Real Effects of Academic Research." *American Economic Review* 79, no. 5 (1989): 957–70.

Jaffe, Adam, Manuel Trajtenberg, and Rebecca Henderson. "Geographic Localization of Knowledge Spillovers as Evidenced by Patent Citations." *Quarterly Journal of Economics* 108, no. 3 (August 1993): 577–98.

Kain, John F. "Housing Segregation, Negro Employment, and Metropolitan Decentralization." *Quarterly Journal of Economics* 82, no. 2 (May 1968): 175–97.

Kane, Yukari Iwatani. "Overseas Start-Ups Move In." *Wall Street Journal*, May 26, 2011.

Kerr, William R. "The Agglomeration of U.S. Ethnic Inventors." In Edward L. Glaeser, ed., *Agglomeration Economics*, pp. 237–76. Chicago: University of Chicago Press, 2010.

Kissack, Andrea. "Electric Vehicle Companies Tap Silicon Valley Cash." *Morning Edition*, NPR, October 13, 2010.

Klepper, Steven. "The Origin and Growth of Industry Clusters: The Making of Silicon Valley and Detroit." *Journal of Urban Economics* 61, no. 1 (January 2010): 15–32.

Kline, Patrick, and Enrico Moretti. "Local Economic Development, Agglomeration Economies and the Big Push: 100 Years of Evidence from the Tennessee Valley Authority." November 2011.

Kraemer, Kenneth L., Greg Linden, and Jason Dedrick. "Capturing Value in Global Networks: Apple's iPad and iPhone." July 2011.

Krueger, Dirk, Fabrizio Perri, Luigi Pistaferri, and Giovanni L. Violante. "Cross-Sectional Facts for Macroeconomists." *Review of Economic Dynamics* 13, no. 1 (January 2010): 1–14.

Kulkarni, Sandeep C., Alison Levin-Rector, Majid Ezzati, and Christopher J. L. Murray. "Falling Behind: Life Expectancy in U.S. Counties from 2000 to 2007 in an International Context." *Population Health Metrics* 9, no. 16 (June 2011).

Lee, Kyungjoon, John S. Brownstein, Richard G. Mills, and Isaac S. Kohane. "Does Collocation Inform the Impact of Collaboration?" *PLoS One* 5, no. 12 (December 2010).

Leuty, Ron. "SF Life Science Hub Lures East Coast Venture Firms." *San Francisco Business Times*, May 27–June 2, 2011.

Lin, Jeffrey. "Technological Adaptation, Cities, and New Work." *Review of Economics and Statistics* 93, no. 2 (May 2011): 554–74.

Lochner, Lance, and Enrico Moretti. "The Effect of Education on Crime: Evidence from Prison Inmates, Arrests and Self-Reports." *American Economic Review* 94, no. 1 (2004): 155–89.

Lohr, Steve. "Silicon Valley Shaped by Technology and Traffic." *New York Times*, December 20, 2007.

Lucas, Robert. "On the Mechanics of Economic Development." *Journal of Monetary Economics* 22, no. 1 (July 1988): 3–42.

Ludwig, Jens, et al. "Neighborhoods, Obesity, and Diabetes—A Randomized Social Experiment." *New England Journal of Medicine* 365 (October 2011): 1509–19.

Lychagin, Sergey, Joris Pinske, Margaret E. Slade, and John Van Reenen. "Spillovers in Space: Does Geography Matter?" NBER Working Paper 16188. National Bureau of Economic Research, July 2010.

Machin, Stephen, Pano Pelkonen, and Kjell G. Salvanes. "Education and Mobility." *Journal of the European Economic Association*, doi: 10.1111/j.1542-4774.2011.01048.x.

Mallaby, Sebastian. *More Money Than God: Hedge Funds and the Making of a New Elite*. New York: Penguin, 2010.

Manacorda, Marco, and Enrico Moretti. "Why Do Most Italian Youths Live with Their Parents? Intergenerational Transfers and Household

Structure." *Journal of the European Economic Association* 4, no. 4 (June 2006): 800–29.

Manyika, James, and Charles Roxburgh. "The Great Transformer: The Impact of the Internet on Economic Growth and Prosperity." McKinsey Global Institute (2011).

Markusen, Ann, Peter Hall, Scott Campbell, and Sabina Deitrick. *The Rise of the Gun Belt: The Military Remapping of Industrial America.* New York: Oxford University Press, 1991.

Mas, Alexandre, and Enrico Moretti. "Peers at Work." *American Economic Review* 99, no. 1 (2009): 112–42.

———. "Racial Bias in the 2008 Presidential Election." *American Economic Review* 99, no. 2 (2009): 323–29.

Mattioli, Dana. "As Kodak Fades, Rochester Develops Other Businesses." *Wall Street Journal*, December 24, 2011.

Mayer, Heike. "Bootstrapping High-Tech: Evidence from Three Emerging High Technology Metropolitan Areas." Metropolitan Economy Initiative no. 10. Brookings Institution, June 2009.

Milligan, Kevin, Enrico Moretti, and Philip Oreopoulos. "Does Education Improve Citizenship? Evidence from the U.S. and the U.K." *Journal of Public Economics* 88, no. 9–10 (2004): 1667–95.

Moretti, Enrico. "Estimating the Social Return to Higher Education: Evidence from Longitudinal and Repeated Cross-Sectional Data." *Journal of Econometrics* 121, no. 1–2 (2004): 175–212.

———. "Local Labor Markets." In Orley Ashenfelter and David Card, eds., *Handbook of Labor Economics*, Vol. 4B, pp. 1237–1313. London: Elsevier, 2011.

———. "Local Multipliers." *American Economic Review* 100, no. 2 (May 2010): 373–77.

———. "Real Wage Inequality." NBER Working Paper 14370. National Bureau of Economic Research, 2008.

———. "Social Learning and Peer Effects in Consumption: Evidence from Movie Sales." *Review of Economic Studies*, forthcoming.

———. "Workers' Education, Spillovers and Productivity: Evidence from Plant-Level Production Functions." *American Economic Review* 94, no. 3 (2004): 656–90.

National Academy of Engineering, Committee on the Offshoring of Engineering. *The Offshoring of Engineering: Facts, Unknowns, and Potential Implications.* Washington, D.C.: National Academies Press, 2009.

Paserman, M. Daniele. "Do High-Skill Immigrants Raise Productivity? Evidence from Israeli Manufacturing Firms, 1990–1999." Discussion Paper 6896. Centre for Economic Policy Research, July 2008.

Pélissié du Rausas, Matthieu, et al. "Internet Matters: The Net's Sweeping Impact on Growth, Jobs and Prosperity." McKinsey Global Institute, May 2011.

Powell, Walter, Kjersten Whittington, and Kelley Packalen. "Organizational and Institutional Genesis: The Emergence of High-Tech Clusters in the Life Sciences." In John F. Padgett and Walter W. Powell, eds., *The Emergence of Organizations and Markets*. Princeton, N.J.: Princeton University Press, forthcoming.

Riddell, Lindsay. "East Bay Schools Aim to Refresh Biotech Labor Pool." *San Francisco Business Times*, April 22–28, 2011.

Samuelson, Kristin. "High-Skill Job Openings Abound." *Chicago Tribune*, October 3, 2011.

Schumpeter, Joseph. *Capitalism, Socialism, and Democracy*. New York: Harper and Brothers, 1942.

Scott, Allen. "Origins and Growth of the Hollywood Motion-Picture Industry: The First Decade." In Pontus Braunerhjelm and Maryann P. Feldman, eds., *Cluster Genesis: Technology-Based Industrial Development*, pp. 17–37. London: Oxford University Press, 2006.

Sieg, Holger, V. Kerry Smith, H. Spencer Banzhaf, and Randy Walsh. "Estimating the General Equilibrium Benefits of Large Changes in Spatially Delineated Public Goods." *International Economic Review* 45, no. 4 (November 2004): 1047–77.

Sorenson, Olav, and Toby E. Stuart. "Syndication Networks and the Spatial Distribution of Venture Capital Investment." *American Journal of Sociology* 106, no. 6 (May 2001): 1546–88.

Tam, Pui-Wing. "Technology Companies Look Beyond Region for New Hires." *Wall Street Journal*, September 23, 2010.

Thiel, Peter. "The End of the Future." *National Review*, October 3, 2011.

Thompson, Peter. "Patent Citations and the Geography of Knowledge Spillovers: Evidence from Inventor- and Examiner-added Citations." *Review of Economics and Statistics* 88, no. 2 (May 2006): 383–88.

Tice, Carol. "Geeks of a Feather." *Washington CEO*, March 2008.

U.S. Bureau of Labor Statistics. "Career Guide to Industries, 2010–2011 Edition." 2011.

———. "Industry Output and Employment Projections to 2018." *Monthly Labor Review*, November 2009.

———. "Occupational Employment Projections to 2018." *Monthly Labor Review*, November 2009.

———. *Occupational Outlook Handbook*. 2010–2011.

U.S. Patent and Trademark Office, Technology Assessment and Forecast Branch. "United States Patent Grants—Number of Grants per 100,000 Population, by Metropolitan Area, 1998."

Van Reenen, John. "The Creation and Capture of Rents: Wages and Innovation in a Panel of U.K. Companies." *Quarterly Journal of Economics* 11, no. 1 (February 1996): 195–226.

Vara, Vauhini. "Clean Tech Arrives, with Limited Payoff." *Wall Street Journal*, January 20, 2011.

———. "Red Flags for Green Energy." *Wall Street Journal*, October 12, 2011.

Wessel, David. "Big U.S. Firms Shift Hiring Abroad." *Wall Street Journal*, April 19, 2011.

———. "The Factory Floor Has a Ceiling on Job Creation." *Wall Street Journal*, January 12, 2012.

Wheeler, Christopher H. "Cities and the Growth of Wages Among Young Workers: Evidence from the NLSY." *Journal of Urban Economics* 60, no. 2 (September 2006): 162–84.

———. "Job Flows and Productivity Dynamics: Evidence from U.S. Manufacturing." *Macroeconomics Dynamics* 11, no. 2 (April 2007): 175–201.

———. "Local Market Scale and the Pattern of Job Changes Among Young Men." *Regional Science and Urban Economics* 38, no. 2 (March 2008): 101–18.

Wilson, William Julius. *The Truly Disadvantaged: The Inner City, the Underclass, and Public Policy*. Chicago: University of Chicago Press, 1987.

Wozniak, Abigail. "Are College Graduates More Responsive to Distant Labor Market Opportunities?" *Journal of Human Resources* 45, no. 4 (Fall 2010): 944–70.

Zucker, Lynne G., and Michael R. Darby. "Capturing Technological Opportunity via Japan's Star Scientists: Evidence from Japanese Firms' Biotech Patents and Products." *Journal of Technology Transfer* 26, no. 1–2 (January 2001): 37–58.

———. "Movement of Star Scientists and Engineers and High-Tech Firm Entry." NBER Working Paper 12172. National Bureau of Economic Research, 2006.

———. "Present at the Biotechnological Revolution: Transformation of Technological Identity for a Large Incumbent Pharmaceutical Firm." *Research Policy* 26, no. 4–5 (December 1997): 429–46.

Zucker, Lynne G., Michael R. Darby, and Jeff S. Armstrong. "Commercializing Knowledge: University Science, Knowledge Capture and Firm Performance in Biotechnology." *Management Science* 48, no. 1 (January 2002): 138–53.

———. "Geographically Localized Knowledge: Spillovers or Markets?" *Economic Inquiry* 36, no. 1 (January 1998): 65–86.

Zucker, Lynne G., Michael R. Darby, and Marilynn B. Brewer. "Intellectual Human Capital and the Birth of U.S. Biotechnology Enterprises." *American Economic Review* 88, no. 1 (March 1998): 290–306.

Zucker, Lynne G., et al. "Minerva Unbound: Knowledge Stocks, Knowledge Flows and New Knowledge Production." *Research Policy* 36, no. 6 (July 2007): 850–63.

INDEX